Exploring with Geometry Expressions in High School Mathematics

Ian Sheppard
Wesley College in Perth Western Australia

Saltire Software, Inc.
Tigard, OR, USA
www.saltire.com
www.geometryexpressions.com

geometry expressions

ISBN 1-882564-12-X

Limited Permission for Reproduction

Saltire Software
P.O. Box 230755
Tigard, OR 97281-0755
http://www.geometryexpressions.com/
http://www.saltire.com/
support@saltire.com

Table of Contents

Chapter 1 – Introduction

New technologies provide exciting opportunities for teachers to enhance student learning. Interactive computer software is one such technology and can be combined with interactive white boards to foster small and large group interaction. Such dynamic interaction provides opportunities for students to access mathematics in new ways and supports a greater range of learning styles. The dynamic interaction provides teachers with an additional component to their repertoire of instructional strategies. Interactivity engages learners and assists in supporting independent student investigations. Using *Geometry Expressions* (GX), for example, students can readily explore concepts, verify numerical solutions to geometry problems and explore properties of geometrical figures. Such activities support students search for patterns and construction of conjectures. These exercises correlate strongly with the process strands of common mathematics standards as well as the underpinnings of algebraic reasoning.

I believe that students should be empowered as much as possible to take control of their own learning. It is through play that conceptual development is often the most rapid and long lasting. Vygotsky observed that children in play displayed intellectual development beyond what they showed in a school environment. (Kozulin 2007) Students engaged in inquiry and discovery activities may well be involved in intellectual play. A feature of the "play" environment is the capacity of the player to manipulate the objects and try things out. It is a safe environment in which to experiment. A strong case can be made for intellectual play, a situation enhanced by the use of appropriate empowering technologies.

Unfortunately many of our high school students have little opportunity to play, particularly in mathematics and particularly where the focus is on the ability to perform for the test. Many students see school as a drudgery to be waited out. Where this occurs it is sad and any way the spark of curiosity and then discovery can be reignited is a blessing. Interactive geometry programs are well known and powerful contributors to student learning. They provide a rich environment for exploration of geometric relationships. GX adds something new to the genre of interactive geometry as the construction process is more intuitive and potentially accessible to many more students.

There are aspects of many interactive geometry programs that are difficult for the novice (student) user. For example creating a construction from scratch often requires a lot of geometrical knowledge which is beyond the student. To allow access to the benefits of Interactive geometry programs students are often presented with pre-prepared sketches and students then play within the limited field these sketches provide. Pre-prepared sketches hide much of the intellectual work involved in creating the drawing in the first place. It seems as if the power of the software is just being tapped rather than unleashed. GX offers a potential pathway to loosen the leash.

The activities in this book are based around GX. To draw upon a culinary metaphor, the book is intended to be a buffet with the teacher picking and choosing a menu that fits the whims and tastes of their class. The activities are wide ranging and touch much of the geometry and trigonometry taught in High School mathematics courses. The ability to solve trigonometry problems in an intuitive drawing / construction mode is exciting as it opens new ways of learning and teaching. When students are able to discover for themselves the ideas and relationships of geometry, learning outcomes particularly in terms of retention and understanding are much improved. Of course, there is no one approach that will work for all students. Some require more hand-holding particularly when working with a new and unfamiliar tool while others need only the briefest of exposures to see how the use of a new tool enables their learning.

With these experiences in mind, the activities in this book involve the use of GX in high school mathematics classes. The materials are designed to be flexible. They may be used as the basis for an independent geometry unit, as a set of labs for reinforcement or enrichment for a text based unit, as individual activities to be woven into the mathematics curriculum on appropriate occasions, as a support for differentiated instruction in the form of self-guided discovery and extension or as the basis for contextualized learning activities.

Content and Organization

Chapters 2 – 4 discuss key features of GX and the pedagogy underpinning the lab activities in chapters 5 to 10. GX is described as an interactive symbolic geometry program. Interactive geometry is a software genre with well known commercial packages, shareware programs and open source offerings. The features of this genre are documented elsewhere. These introductory chapters focus on *Constraints* (Chapter 2) and *Symbolics* (Chapter 3), features that extend interactive geometry and a scenario (Chapter 4). The scenario is a passionate appeal for the use of an investigative approach with the teacher becoming more of a guide on the side.

Chapters 5 – 10 are the lab activities. They are organized in themes which are likely to be covered in a number of high school mathematics courses. These themes are content based and develop from Middle school or beginning high school through to topics in advanced placement courses. For example the chapter on trigonometry begins with the Pythagorean Theorem and extends through right triangle trigonometry, the sine formula, law of cosines and polar coordinates. The chapter on locus begins with the circle, and ends with Bezier curves. Each of these chapters is prefaced with an outline of the labs. Each lab has teacher notes to assist in planning.

The themes and the chapters are:

- Chapter 5 Congruence and Similarity
- Chapter 6 Proof
- Chapter 7 Transformations
- Chapter 8 Trigonometry
- Chapter 9 Coordinate Geometry
- Chapter 10 Loci

While the labs are self-contained, the development within each chapter assumes a growing knowledge of GX and ability to construct the drawings. Consequently initial labs have quite detailed instructions while later labs have less detail, assuming that students have prior experience with the drawing tools required. This is consistent with encouraging students to become increasingly independent and to use the software to support their own learning. Consequently some preliminary instruction may be required for advanced classes beginning with later activities within the chapters. Using interactive white boards has been found to be very effective in demonstrating the drawing tools in GX.

The student manual (in press) provides a subset of the activities in this book associated with a particular course of study.

The activities in each unit provide a framework for students' explorations. Often there are one or two activities requiring very little technical expertise that you will feel comfortable giving your students with little or no guidance beforehand. Other activities maximize guided discovery by students. Some of these are more open-ended than others, and their difficulty levels vary. You should choose those activities that are most appropriate for the ability and skill levels of your students. The text provides blackline masters for student work.

Terminology:

GX can be used with menu options or icons. The icons arranged in toolboxes duplicate most of the menu options. The icons are shown sometimes, often the first time a tool is used in the text and the menu item or description used thereafter. The menu item is referred to using the vertical bar to separate levels. For example
Edit | Settings | Math | Math | Angle Mode means

- choose the Edit menu,
- choose Settings from the Edit menu,
- choose the Math tab in the left sidebar
- open ⊞ → ⊟ the Math window
- choose Angle Mode from the Math window.

Draw | Line Segment is the  icon from the Draw toolbox.

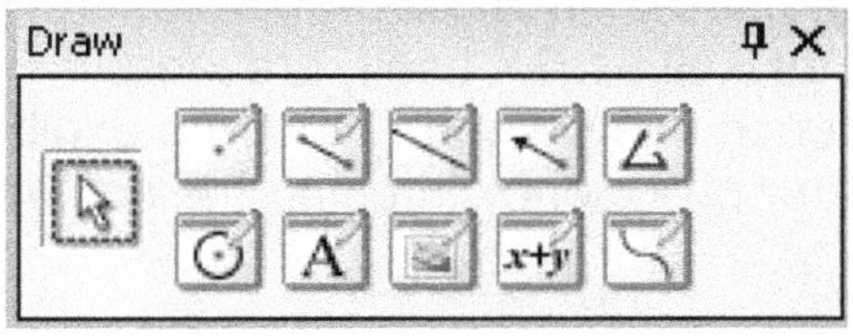

The first time a particular construction or program feature is used a detailed description is given. A statement of the construction is made and then the steps detailed in point form. Where the feature is used subsequently the point form detail is not provided.

For example in an early Lab:

Construct a line segment of length 4.

> *From the Draw menu select the Line Segment tool* .
> *Click, drag and release is one way to draw the line segment.*
>
> *Choose the Select tool* .
> *Click on the line segment.*
>
> *Click the Constrain Distance/Length tool* 
> *Type in 4 and press Enter.*

Subsequently the Activity might just say
Construct a side of length 4.

Hints and suggestions

If you're new to GX, work the activity yourself in your preparation for working with students. Becoming familiar with the activity can greatly facilitate your own learning of the software and increase your comfort level with these materials in the classroom. *GX*'s online help is useful for clarifying the use of a menu or tool and in understanding GX specific language.

Even if you're using GX for the first time, don't let your inexperience keep you from placing the software into the hands of your students! One of the best aspects of using interactive geometry software is learning along with the students. Invariably when I have had a problem in class it is one of the students who is able to find a solution, teaching me and the rest of the class. Such experiences are valuable in modeling approaches to learning, moving the dynamics of the classroom to a more collaborative learning environment and valuing the skills students bring to the classroom. Nothing in the classroom can compare to the excitement generated when students and teacher experience the joy of learning mathematics together.

Where an activity has an extension component, there is an opportunity for more open investigation. For the able students this is where the most valuable learning can be gained.

Regardless of the degree of structure you are comfortable with, be sure to allow time for students to simply explore with the software. Before or after initial activities students enjoy trying the tools and experimenting on their own. This free exploration motivates students to want to learn more about the capabilities of the software and invariably leads to mathematical discovery or consolidation. Where a student holds a geometrical misconception this experimentation can provide the crucial discord required for the student to re-conceptualize and thus move forward with their learning in mathematics.

I encourage you to allow your students to work in pairs or in small groups for some if not most of the activities. The units in this book, as well as GX itself, are ideal instruments for cooperative learning. When we teach this way, our students exhibit both increased mathematical understanding and greater proficiency in using the software.

Chapter 2 – Constraints

Constraints is a feature distinguishing GX from other dynamic geometry programs. This chapter briefly describes attributes in drawings that can be constrained, an application using *Constraints* as a pedagogical approach compared to other interactive geometry programs, the use of *Constraints* to solve a problem and to develop understanding of a problem. These examples indicate reasons why *Constraints* have application in high school mathematics courses.

Think of *Constraints* as a measurement that we fix in our drawing. For example, when drawing or constructing a figure on paper using drawing instruments we draw lines of particular lengths (using ruler and compass) and create angles of particular measure (using a protractor or template). Constraints mirror this process in an electronic construction. We can draw a line and then set its length or select intersecting lines and set the angle. There are many more attributes that can be constrained,

In addition GX allows the user to constrain attributes such as coordinates of points, perpendicularity, parallel, direction or slope of lines, vectors, circles, coincidence, tangency, congruence, equation and point proportional. The toolbox to the right illustrates the range of properties that can be constrained.

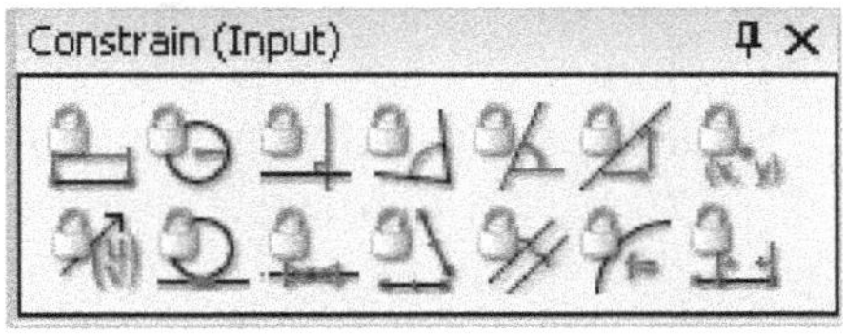

Using *Constraints* is closely aligned with traditional construction techniques. Consequently I find students are more able to create drawings independently of the teacher than with other dynamic geometry software. The following example involves exploring the converse to the Pythagorean Theorem and compares the use of a non-constraint based interactive geometry system with the use of *Constraints.*

Explore the converse of the Pythagorean Theorem:

How do we know whether or not a triangle has a right angle given the lengths of each side? A teacher might begin by posing the question, "Is a triangle of sides 5, 6 and 7 right angled?" Clearly drawing an accurate sketch will give the student a feeling for whether or not this is the case with a follow up using the theorem.

To draw a sketch using dynamic geometry software I would ask the students to

- draw a triangle,
- measure the lengths
- drag vertices until they get a triangle close to the desired measurements.

The diagram on the right shows such a sketch.

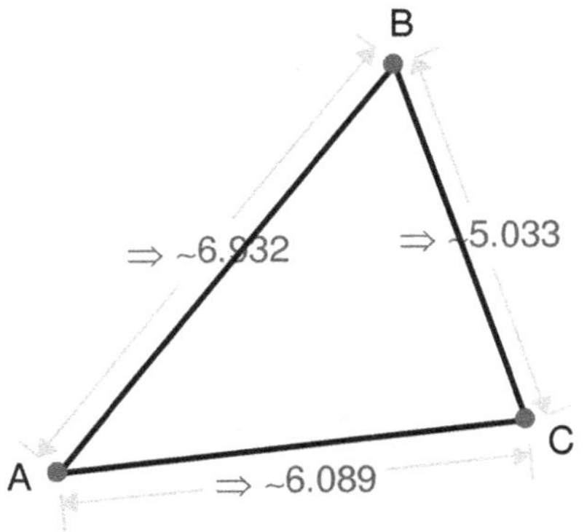

The sketch can be used to explore other side lengths by dragging the vertices to change the side lengths.

Note that the side lengths will not fit the original problem specifications exactly but it is close enough to be convincing that there is no right angle in a 5 – 6 – 7 triangle. Dynamic Geometry software enables this exploration to be done readily and intuitively by students.

However, if the triangle is has an angle close to or equal to 90° we are left with a dilemma: How can we be sure if the triangle is right angled? To answer this, an accurate drawing is required. The table below demonstrates the ease with which the use of *Constraints* enables students to answer the question and to appreciate the underlying theory.

To draw an accurate (exact) sketch using dynamic geometry software

With *Constraints* (GX).	No *Constraints*.
• draw the triangle, • select one side and constrain to length 5 • select a second side and constraint to length 6 • select the third side and constraint to length 7 • observe the angles and if required measure them.	Apply the same construction techniques as for a pencil and paper construction. That is • draw a point • translate it 5 units • draw the line segment • translate the original point 6 units • draw a circle of radius 6 • translate the other end of the line segment 7 units • draw a circle of radius 7 units • mark the intersection • connect to the end points of the original line segment • hide the construction lines. • observe the angles and if required measure the angles
Not only are we able to focus on the presence or otherwise of a right angle in the triangle it is easy to use the same drawing and change the *Constraints* to explore triangles with different side lengths.	Perhaps many students will have forgotten what the original purpose is by now. Yes we can remind them and there is wonderful geometry involved, however it is not the focus of the activity. Creating such a diagram independently is I suggest limited to the very best students and is beyond the overwhelming majority of high school students.

Using *Constraints* for such an investigation allows students to focus on the mathematics that is the point of the lesson, in this case whether the triangle is right angled. Using *Constraints* is fast, easy to explore multiple cases and enables students to work with the essential geometric facts associated with the problem.

Note: Construction techniques such as those outlined using "*No Constraints*" interactive geometry software in the above example are used in activities in this book where the construction supports the main purpose of the lesson such as developing a proof.

With geometry it doesn't take much for me to be confronted with challenging problems and to experience the frustrations associated with being stuck. It is a healthy reminder of what we ask of high school students. The following problem was given to me by one of my students. I got stuck but was able to make headway using *Constraints*. I was able to use GX *Constraints* to create a drawing very quickly and improve my "feel" and understanding for the problem.

In triangle ABC, AB = 8, BC = 7, and AC = 5.

We extend AC past A and mark point D on the extension.

The bisector of $\angle DAB$ meets the circumcircle of ΔABC again at E.

We draw a line through E perpendicular to AB. This line meets AB at point F.

Find the length of AF.

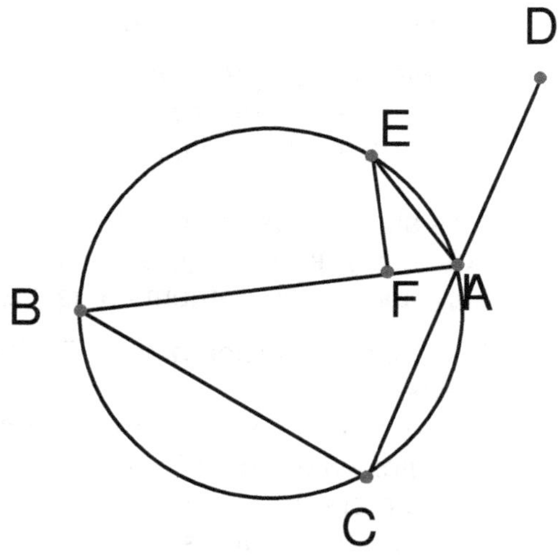

At this stage if you don't have lessons to prepare for tomorrow, stop reading and see what progress you can make with the problem.

I got stuck. It felt like an accessible problem. I tried constructing some extra lines, there is a circle, angles in the same segment etc. but then I need some angles. Maybe there is something special about the triangle, clearly not right angled and so not a diameter of the circle. Hmmm. I can't even look up the answer as the problem came from a student.

Eventually I turned to GX to construct the diagram using *Constraints*. This is how I did it

Draw a circle

Draw three points on the circle
- Hide the center of the circle
  - Select the point and right click
  - Select Hide from the Context menu

- Re-label the points to match the problem

Draw line segments and constrain the lengths for the sides of the triangle.

Extend AC past A
- Draw the infinite line AC
- Draw a point on the line and change the label to D
- Hide the infinite line
  - Select the line and right-click the line
  - Select Hide from the Context menu
- Draw the line segment AD

The bisector of *∠DAB*

- Construct | Angle bisector

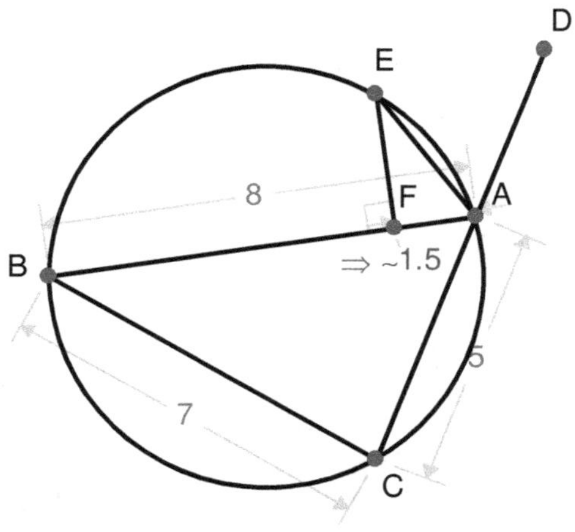

Constraints

Mark the intersection of the Angle bisector and the circle at point E.

Draw a line segment from E to AB

Constrain the angle EFB to a right angle.

Find the length of AF.
Select points A and F
Calculate | Real |Distance/Length.

This sketch was made with GX and AF is calculated to be 1.5. Why? The answer was not satisfying as there is no indication of why. What is it that is special about these particular measurements or is it a general property that is captured in this problem?

Measuring different things $\angle BAC$ or $\angle EAD$ is 60°. This can be checked with the law of cosines.
EA bisects the exterior angle so they are also 60° angles and ΔAEF is a 30° – 60° – 90° triangle. I want to find why AE = 3.
Some more measuring reveals BE = 7 and CE = 7 so ΔEBC is equilateral. Now I have made some progress and the outline solution below makes use of those insights provided by calculating using GX.

Outline for solution:

$8^2 = 7^2 + 5^2 - 2 \cdot 8 \cdot 5\cos\angle BAC$	Law of cosines
$49 = 64 + 25 - 80 \cos\angle BAC$	
$-40 = -80 \cos \angle BAC$	
$\cos \angle BAC = 0.5$	
$m\angle BAC = 60°$	
$m\angle BAD = 120°$	Supplementary to BAC
$m\angle BAE = 60°$	Bisected angle is half measure
$m\angle EAB = m\angle ECB = 60°$	Angles in same segment
$m\angle EAC + m\angle EBC = 180°$	Opposite angles of a cyclic quadrilateral
$m\angle EBC = 60°$	$EAC = EAB + BAC = 60° + 60° = 120°$
ΔEBC	Angles are 60°
$BE = CE = 7$	Equilateral triangle
$AB \cdot CE = AC \cdot BE + AE \cdot CB$	Applying Ptolemy's theorem to ACBE
$8 \cdot 7 = 5 \cdot 7 + AE \cdot 7$	
$AE = 3$	
$AF = AE \cos 60°$	Right triangle trig
$= 3 \cdot 0.5$	
$=1.5$	

I found my learning from the exploration was more lasting than reading a solution. This may reflect my study skills. Reading a solution I, and I suspect many others, see this as the end point and learning stops. I might appreciate the elegance and insights of a written solution and intend to remember. The reality is I haven't engaged sufficiently with the problem and solution to recall it later.

So many students do this when preparing for exams and tests believing they are learning. Later they are confronted with a similar problem, possibly the exam, and experience the same difficulty; that is they haven't learned sufficiently from seeing the solution to solve a similar problem. I would say to students "be determined to know it. Don't let the examiner catch you out with that question again. Make sure you really learn it." Students rarely do!

While teachers are fully engaged with the day to day life of working in a school some do create their own problems to fit their particular circumstances. One might wonder what was the inspiration or ideas which lead to the creation of the previous problem. The main ideas could be the two different 60° triangles with integer sides or an equilateral triangle inscribed in a circle. Tools such as GX can bring to light new insights about the geometry which may lead to creating problems for our students. Invariably those problems we adapt or develop ourselves lead to greater student engagement as we put more of ourselves into our work.

Using *Constraints* is an intuitive way to create drawings, provide opportunities to solve difficult problems and to experiment with figures. When combined with *Symbolics*, *Constraints* allow to and fro movement between the specific example and generalization as is discussed in the next chapter.

Chapter 3 – Algebra in GX (*Symbolics*)

This chapter provides a background of *Symbolics* in the context of geometry. The pedagogic value of integrating *Symbolics* into the teaching of geometry is discussed and exemplified.

The development of algebraic representations for patterns is an important aim for High School mathematics courses. The National Council of Teachers of Mathematics (NCTM) has produced widely accepted standards for school mathematics. These standards include geometry and algebra in the content strands and problem solving, reasoning and proof, representation and connections in the process strands. Specifically for High School, Standard 8 is *Geometry from an Algebraic Perspective* (NCTM 1989).

In 2006, the Mathematical Association of America (MAA) launched a National Science Foundation (NSF)-funded effort to review existing research on the teaching of algebra across K-16, to suggest appropriate content and pedagogy for the future K-16 algebra curriculum. The report is *Algebra in the 21st Century Curriculum* (Katz 2007). Bob Kansky's (2007) summary states:

> *Early algebra should focus on two central features: developing conceptual understanding of operations and relationships through "generalizing arithmetic;" and practicing important skills and procedures in meaningful contexts. Algebra at these levels [introductory algebra and intermediate algebra] should focus on underlying ideas such as algebraic expression, equation, **connections** among algebraic representations (verbal description, table, graph, … As a goal of this refocusing, students should see mathematics as "useful in solving problems."*
>
> *Determine the impact on student learning of a refocused college algebra course. The typical college algebra course focuses on algebraic manipulation; … For the most part, instruction in college algebra is ineffective. … Both MAA and the American Mathematical Association of Two-Year Colleges have proposed the development of college algebra courses "in which students address problems represented as real world situations by creating and interpreting mathematical models"* (*np*).

The focus in elementary school is on recognition and word explanations for number patterns. In parts of the world this is referred to as pre-algebra and is supporting algebra as the language of generalization. The development of number sentences leading towards simple equations follows without any formal algebraic representation. Recent changes to curricula have moved the beginnings of formal algebraic representation and manipulation into the middle school years. This is represented in topics like the coordinate plane and linear equations.

Symbolics in GX is the formal algebraic representation of patterns in geometric figures and relationships. This is most familiar when the geometry is situated in the Cartesian plane. Drawing and interpreting graphs of functions will be most students' first experience of the multiple representations possible, representations such as a graph or visual image and the algebraic or functional representation. It is also to be found in transformations on the Cartesian plane such as describing a reflection in the y-axis. Geometry is a rich environment for strengthening students' understanding of symbolic representations beyond a number pattern, an equation and graphs of functions.

Working with variables in GX

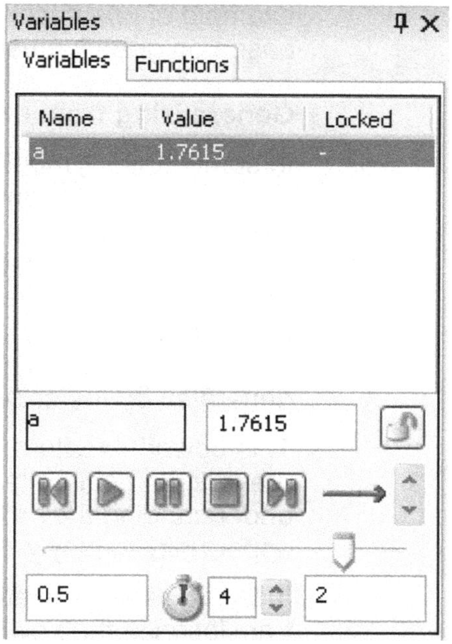

Constraints can be numbers or variables. GX supplies variables as the default option when creating a constraint. The variable window displays the current variables and their values. Not only is the value displayed but the value of the variable can also be changed. A value can be entered or changed dynamically using the slider. Multiple representations and dynamic displays are valuable in supporting robust and flexible concept development.

Creating an animation requires working with variables and is controlled from the Variables window. Controlling the animation involves determining the appropriate range of values for the variable. As the animation runs the variables value is seen to change dynamically as the figure moves. This allows students to conceptualize symbols as variables, a crucial component of algebraic thinking. I have come across many high school students who believe that a symbol represents a specific value only. These students have probably extensive experience solving equations and have come to see the symbol as representing a specific number.

The idea of substitution is explicitly displayed in the variable window.

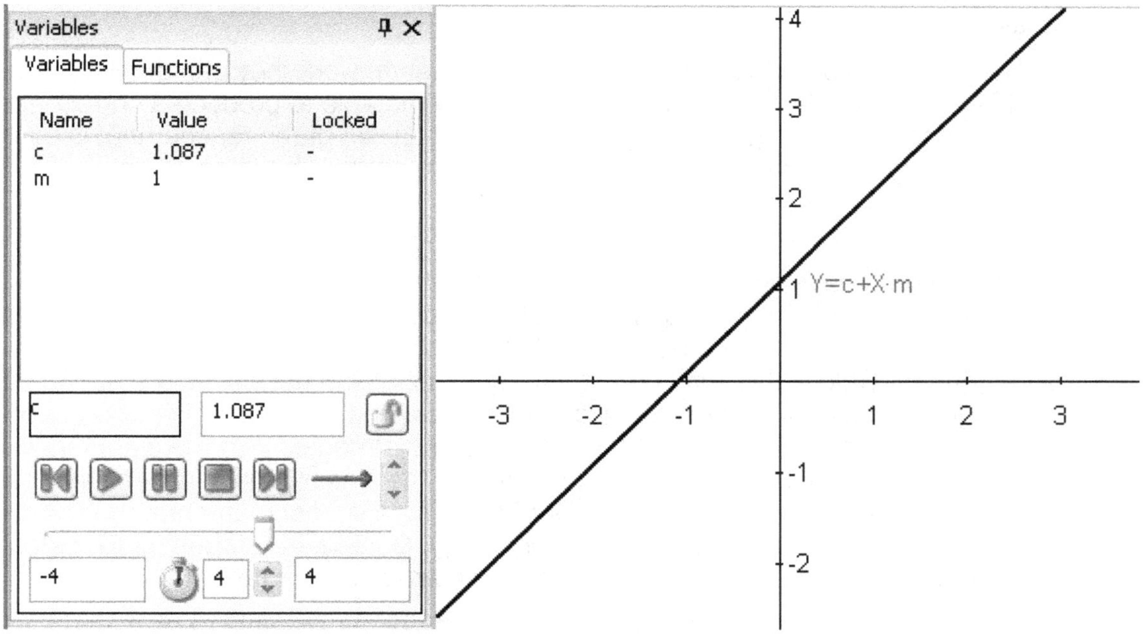

The function was entered as $y = mx + c$. The function's parameters appear in the variable window. The slider can be used to change the value and the line moves dynamically. From here students might:

- Change the value of *c* and watch the graph slide up and down the axis;
- Change *m* and observe how the slope changes and y-intercept remains fixed;
- Add a real measurement of the slope to enhance the link between *m* and slope
- Play it as an animation.

The example illustrates how GX has function display capabilities. Appendix A is an example of one such lab activity. Activities drawn from other sources such as those based on graphing using graphing calculators can be readily adapted for use with GX.

Generalizing from examples

In some labs, *Symbolics* are mainly used to sum up the investigation and encourage description of the generalizations in symbolic form. For example, Lab 16, *Sliding with Coordinates* explores translations in the Cartesian plane. The guided investigation leads students to describe translations in terms of vectors. Step by step the generalization is developed as a pattern, described in words and finally symbolically as transformation equations. As the *Constraints* can be set to symbols, the algebraic representation of the effects of a translation on the coordinates of a point can be generated by the software.

This powerful feature can be used to enhance student learning. However it can also prevent the desired learning occurring. I believe the key difference relies on the teacher understanding their students. It is a case of ensuring that students have had sufficient opportunity to play with the raw material from which they are to generalize.

In the translation example the labs are structured to encourage students to describe translation in their own words and to create animations involving translation. This is the play period, a time when students have the opportunity to become familiar with the terms, to manipulate objects and to describe translations in their own words. It is only when the students are familiar with the transformation and have developed the capacity to appreciate the simplicity of the symbolic description that Lab 3 will be a worthwhile activity.

Should students do Lab 3 without that developmental work they can get the answers and learn nothing that will assist them beyond being able to produce a correct answer to an almost identical question. It is the professionalism of the teacher that is essential in planning when such tools are going to enhance student learning.

Technology and proof

In my classes I have times when there are no calculators, times when calculators may be used and times when calculators must be used. No calculators means I want my students to practice their number skills, the focus is on efficiently and accurately using an algorithm. Calculators may be used means the lesson objectives are not related to the performance of number calculation algorithms. The calculation is merely supporting working with another concept and the student who struggles with the algorithm will focus on the process and not on the objective of the lesson. Using a calculator can enable such students to participate more fully in high school mathematics. The same argument applies to many aspects of technology. Consider a problem such as proving that the diagonals of a rhombus bisect each other at right angles.

In my typical class tackling this problem, some students are able to do it unaided, some with assistance and others seem unable to access all the different components required at the same time. For these students, with persistence they are often able to reproduce the individual components of the proof but are unable to see why or appreciate the compete story. Using GX can make this story clearer as the whole story is mapped out without the additional cognitive load associated with so much algebraic manipulation.

For a coordinate geometry proof each step of the proof, excluding some algebraic manipulation and rearrangement, can be duplicated in GX. The essential steps are calculations based on the figure. These are shown in the following screen shot.

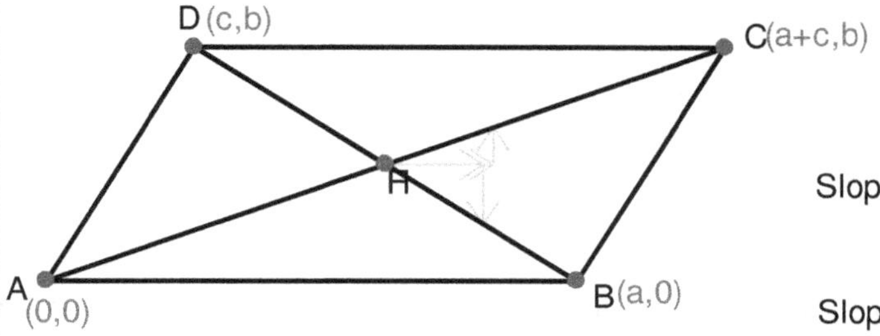

Slope AC $z_1 \Rightarrow \frac{b}{a+c}$

Slope BD $z_0 \Rightarrow \frac{-b}{a-c}$

Midpoint AC $\Rightarrow \left(\frac{a+c}{2}, \frac{b}{2}\right)$

Midpoint BD $\Rightarrow \left(\frac{a+c}{2}, \frac{b}{2}\right)$

product of slopes $z_0 \cdot z_1 \Rightarrow \frac{-b^2}{(a+c)\cdot(a-c)}$

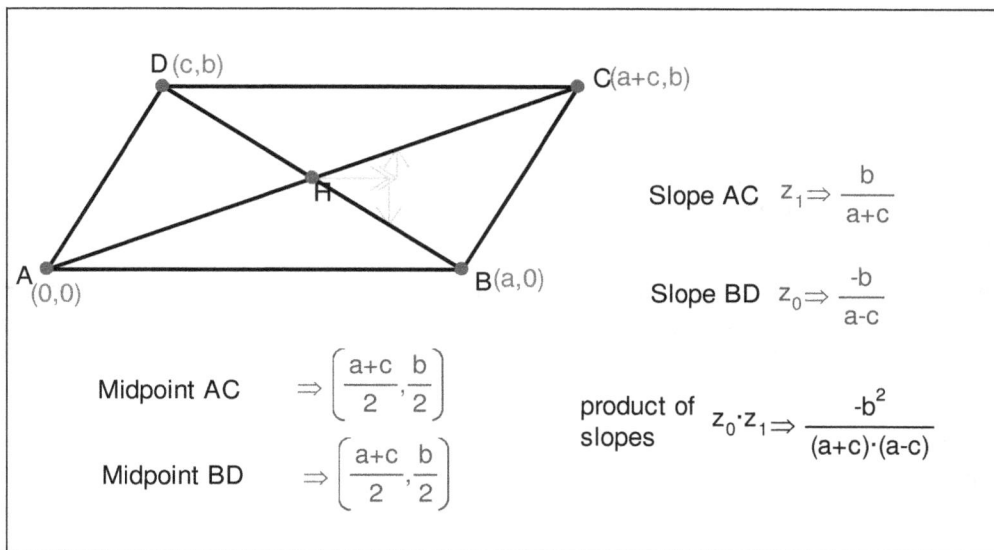

Doing the Calculations in GX enables students to appreciate the steps in developing the solution. Being able to see a bigger picture can help students with such problems. is not difficult to imagine a student who is lacking familiarity and or confidence in the application of the midpoint formula to focus on that aspect of the calculations. In the process they are likely to have lost sight of the goal and to integrate the steps into a proof.

The proof follows with the GX calculations shown in bold type.

Proof:

Without Loss of Generality,

Consider a rhombus with vertices at A(0, 0), B(*a*, 0), C(*c*, *b*) and D(*a* + *c*, *b*)

Midpoint of AC is $\left(\frac{a+c}{2}, \frac{b}{2}\right)$

Midpoint of BD is $\left(\frac{a+c}{2}, \frac{b}{2}\right)$

AC and BD have the same midpoint. Therefore AC and BD bisect each other.

Slope of AC = $\frac{b}{a+c}$

Slope of BD = $\frac{-b}{a-c}$

Slope of AC × slope of BD $= \frac{b}{a+c} \times \frac{-b}{a-c}$

$$= \frac{-b^2}{(a+c)(a-c)}$$

$$= \frac{-b^2}{a^2-c^2}$$

$$= \frac{-b^2}{b^2}, \ (b^2 + c^2 = a^2) \text{ as AD = AB}$$
$$= -1$$

AC $\perp$ BD

The student still is required to do some algebra to complete the proof. The slopes of the diagonals of the rhombus have been calculated symbolically.

Working with geometric figures in the coordinate plane can assist students to focus on the mathematics central to the lesson, in this case the proof, and be less caught up in the algebra. When the mechanical processes involved in the construction dominate, there are many students who miss out on appreciating the key points of the argument.

Discovery

Most discovery activities with Interactive geometry use measurement to develop conjectures. For example students are able to readily conjecture that the angles in the same segment are equal by creating an appropriate diagram, measuring the subtended angle and dragging the point; *Circle theorems* Lab #13. *Symbolics* can support discovery through observations of expressions.

For example the lab *Centers of a triangle* provides starting points for many investigations. From the triangle incircle, Area of the triangle and radius can be calculated.

The drawing shows a triangle with its incircle drawn and expressions displayed for incircle radius and triangle area.

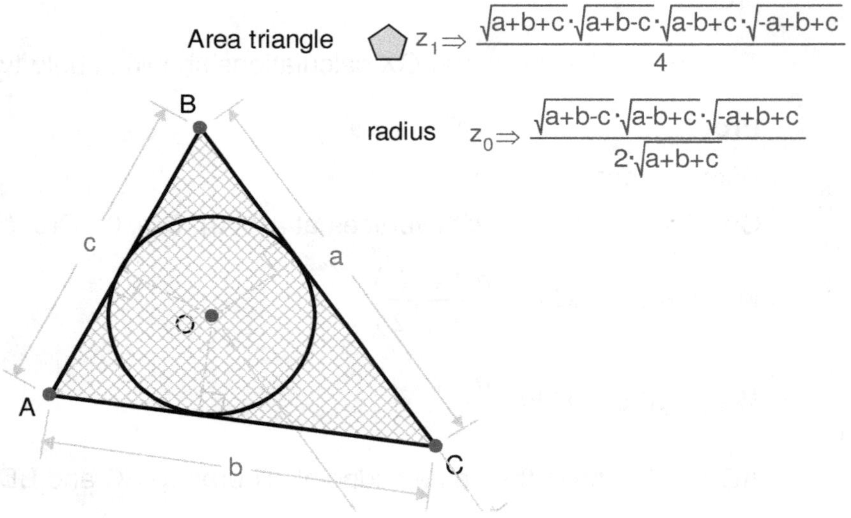

The similarity between the two expressions suggests a simpler relationship between radius and Area. The ratio of Area to radius is $\frac{1}{2}(a+b+c)$ or half the perimeter of the triangle. Rearranged this becomes $r = \frac{2A}{P}$. Such an elegant relationship, what about proof?

The Area of ΔABC =|ABC|
= |AOC| + |AOB| + |BOC|

$$= \frac{1}{2}br + \frac{1}{2}cr + \frac{1}{2}ar$$

$$= \frac{r}{2}(a + b + c)$$

$$= \frac{rP}{2}$$

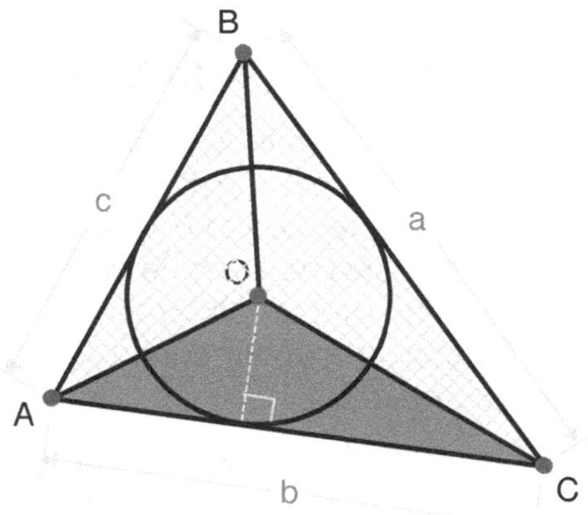

Creating problems

A calculus course I was teaching was in its first year and included Bezier curves. There were insufficient problems for students to practice and I wanted to make some up. Being able to drag a figure to the shape I wanted and then display the equations meant that I could create my own questions. I would look at the equations modify the coefficients to “nicer” numbers, plot to check it was still the desired shape and then build the calculus part of the question. I was making use of technology to assist me in lesson preparation.

Arguments about the use of technology in High school Mathematics courses are not new. Not so long ago the use of calculators for high school mathematics tests was banned. If students were unable to use them for the test they were unlikely to be used in many classrooms to support learning. Now, there are many courses where calculator use is encouraged and school districts such as NY City have mandated that students have graphing calculators available when sitting the exams. That is, it is recognized that technology has its place as it is argued that students are able to think about the material being taught or examined rather than concentrating on the mechanics of number calculations. That real world data or numbers can be used just as easily as data contrived for ease of calculation is an added advantage.

Working with Expressions

GX supports calculating with expressions. In the coordinate geometry proof the product of the slopes is calculated by GX. This capability of GX is used sparingly in this book, however it does have great potential for more advanced students. The paper “*Insight with Geometry Expressions*” is printed as Appendix B and illustrates the possibilities for more advanced applications and integration with computer algebra systems (CAS).

Chapter 4 – Danny's room

Danny's room is a diarized account of teachers starting to use GX in the classroom. Danny is a composite, enthusiastic and passionate high school mathematics teacher. He is introduced to GX at an NCTM meeting and suitably enthused begins to use it in his classroom. Being a composite character enables Danny to illustrate positive learning outcomes for students at different levels and points to ways in which a practicing teacher can incorporate GX into everyday practice, including the inevitable frustrations.

> *Participants will be engaged in a wide range of hands on activities that will provide action and discussion on representation of teachers as problem solvers or algorithmists, pedants or creative human beings. A package of materials and activities will be readily adaptable for use in the classroom* (Frossinakis and Sheppard 2007).

This is the promotional blurb for a workshop I attended at NCTM's Annual meeting. It was quite a lively workshop with the presenters acted as a couple of farmers. They chose the role play to contextualize the workshop and maximize the area problem. They used this problem to illustrate multiple approaches to problem solving and uses of technology.

Alpaca problem:

The *Constraints*:

- Fencing 60 panels each10' long
- 1000 square feet per alpaca
- Build alongside existing fence
- Maximum alpaca paddock area

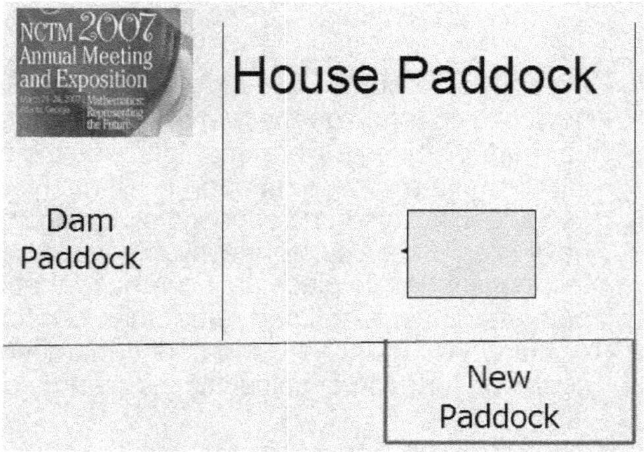

We worked on the problem in groups. In my group we developed two approaches, a numerical approach using a graphing calculator and an algebraic approach. To finish off the presenters showed some other approaches using technology.

The problem was set up in *Geometer's Sketchpad* (*GSP*). Dynamically dragging a corner of the field enabled us to see where the optimum solution lay. Then they raised the question, can we do better if it is not a rectangle? All of the approaches we had used fell apart as there were now too many variables.

"What if we divide the fence into four equal sections of 15 panels each?" The presenter then drew this diagram using GX.

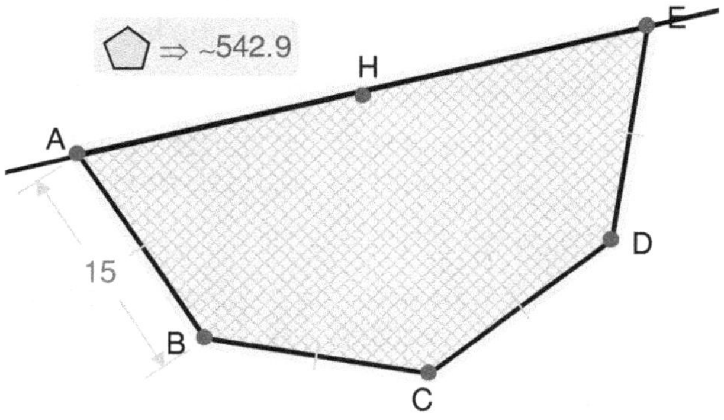

It was drawn so quickly, it felt like only a minute.

The four sections are all of length 15 (15 panels). We can now drag corners and the area is displayed dynamically. And we can use more segments if we like.

Wow!

What excited me was how quickly and easily the diagram was created. The ability to specify the side length seemed so natural in terms of drawing, after all isn't that how we would go about drawing scale diagrams. I could imagine my students enjoying this. OK I have to try this out.

After the session I spoke to the presenters and they were happy to share some GX resources with me. They also pointed out that I could download the software and use it as a trial version for 30 days. On the way home I was planning how I might use this in my Trigonometry class where I was about to start the sine formula and law of cosines.

In previous years most of my students have been able to apply the laws while we do the topic. Retention has not been that good and they struggle with the manipulation, particularly finding the angle using the law of cosines. If I just use GX they will answer the problems OK but they need to be able to do the problems on paper for the State test. Perhaps we can use GX to check their answers.

All the problems lend themselves to the application of *Constraints*, *i.e.* draw a triangle, constrain the known sides and angles and then measure the unknown. Surely this is something my students will be able to pick up and do readily. (As if that ever happens!) Yes I knew it would take a little time for students to develop familiarity with the software.

The conference presenter was happy to share an instruction sheet which I was able to modify for this unit. It was one of those exciting times when I was really looking forward to the lesson.

Next problem was getting the software onto the computers in the lab. Unfortunately we have limited resources to support the computers and Brad is always busy. OK if you want to do that you can but I can't install it till the break. Well can I do it? Sure here is the administrator password. It took me two hours after school and I booked the computer room.

Arriving back at school I brought my laptop to class and connected it to the projector. Hey guys, look at this cool computer program I came across at the conference. *What? You do you do math out of school, sir?* Anyway after the banter we got started and I set out to explain the problems I expected them to be able to do at the end of the topic. *We are going to do one of these now.*

> *Tiger Woods has just hit his drive 324 yards. The hole was 478 yards but the hole has a dogleg so Tiger's drive goes off 22° to the right. How far is his second shot going to be?*

I did a rough sketch on the board. Then I turned on the projector and opened GX on my laptop and passed it to Daniella, one of the students.

OK so we have a triangle with no right angles and we want to know this length.

OK Daniela, see those buttons on the right, click on the Line Segment tool, the one with two dots and a line between them.

This one?

Yeah. Now move the mouse over to the drawing area, yes. You are going to click and drag to draw a line.

Daniela clicks and releases.

You need to hold the mouse button down while you drag. Soon we have a line. *Good, now from one of the endpoints draw another line, move to the endpoint click drag to a new position and then release. Now draw another line to complete the triangle.*

We have a triangle but it doesn't look anything like our diagram so lets put the measurements on it. Danielle moves the mouse over one line and click. Oops, we need to use the Select tool first, the one with the arrow. Daniela finds it and clicks. *Now click on the line again. On the right there is the Constraint tools, see how the first button is colored now, that means we can use it. Move the mouse over it and …* Daniela clicks it and the constraint window pops up. *Now enter the value 324 and press the enter key.*

By the way you can move the figure around. Try dragging the points around now Daniela. See how you can change the shape. In the problem another side was 478. How can you set another side to 478? Just click on the side to select it and enter its length. Try it Daniela. Now when you drag it around those sides don't change in length. So what other information is given in the problem? The 22° sir. Let's specify the angle now. Any ideas how we might do that? Silence, which is pretty unusual as someone is usually distracted but I actually had everyone's attention for the moment, the value of novelty perhaps?

Well click on one line. Good, Now hold the down the ctrl key and click on the other ray for the angle. See in the Constraints toolbox the Angle icon is colored now. Yes click on it and enter 22. Now try to change the shape. Daniela clicks on points and can rotate and drag the shape but it remains the same. *So now we have defined the triangle, how long is Tiger's shot to the hole?*

If I use a ruler, Ok Daniela, click on the line segment we want to measure. See in the Tools panel there is another toolbox called Calculate (Output) and the Distance/Length icon is colored. Before you click on it just click Real. Just above the button there. Ok and yeah that is the distance to the hole. Problem solved.

Everyone should be able to do that.

However in the State exam you need to do this with pen and paper so we still need to learn the theory and how to apply it. The program is just using the same rules that we are about to go through. I want you to use it to check your work.

I felt that the students enjoyed the lesson. It had a bit of the wow factor, the students seemed keen to use the software and I expected they would be able to do so easily.

It was a week later before I could get to the computer lab. The students had made good progress with the sine formula, at least they were solving the problems proficiently. With any new software there is always some familiarization time required. So for the first hands-on lesson I planned a gentle introduction, just checking the home work problems and then doing a few other exercises from the textbook. I thought it would be quick and an incentive.

The class wasn't anything like I imagined. I had this idyllic picture of the students flowing into it and completing lots of problems in an engaged manner. The first part of the lesson I was inundated with questions, it was as if our earlier lesson had left no impact. "Sir it doesn't work was so common." Midway through the class I paused and took a little while to look around. There were students who were competent amongst the group (Danielle and her friends for example) and they were much better at helping their classmates than I seemed to be. Instead of trying to sort the problems myself I began asking the competent students to help others with questions. The atmosphere in the class changed from disaffection to busyness.

I have a reticent group within the class who keep to themselves. By the end of the period they had a sense of achievement too. It came from Dylan watching Connor succeeding and then working with Connor for a while. The interaction was surprisingly cooperative given the normal class dynamics and then Dylan went back to my reticent group of students and began to teach them.

I needed to step back and give the students' time to play, to get a little frustrated and to help each other rather than seeing myself as the fount of all wisdom in the class. This is something I still struggle with but thinking back to this lesson reminds me.

Our next computer lesson was very different. There was no need for my assistance with using GX. They were all away in no time and so I spent my time asking the students about their understanding and did using GX make any difference. The change in routine was appreciated almost universally but there were other comments too.

Abe said that he now understood what the sketches meant. This puzzled me and digging deeper I discovered that he really didn't have an appreciation of angle size apart from quarter and half turns. Now he was drawing more realistic diagrams.

Danielle discovered the law of cosines came out when she accepted the default values for *Constraints*. I asked her to share what she had found with the class. The amazing thing was that the law became some how real as a result. When pressed the closest they could say was that they understood it was a formula applying to the triangle rather than a procedure to be followed.

With this insight we worked through the process of finding the angle using the law of cosines using *Symbolics* in GX. We were able to move between the general and specific by changing the *Constraints* and while doing this Donasia noticed the Variable window and pointed out how we could change the values there and also use the slider to change the values. My students were teaching me.

Well, the test comes with the assessment. We do a common test across the classes and I am pleased to say that the class did comparatively better in this. Not statistically significant but enough to be noticed. Other teachers wanted to know why. I shared what I had done at a department meeting. Now our department head has bought a school license for GX and encouraged us to do more.

When we were reviewing for the finals I did notice that this topic required only the gentlest of reminders. I guess that the retention has been much improved over what I used to do.

Chapter 5 – Congruence

Teacher Notes

Context for this chapter

ALPINE GLASS
Crystal Clear Custom Windows

Alpine Glass specializes in those difficult to make windows.

Aspen Lodge has asked for their premium double glazed panels for their unique architect designed ski lodge. The triangular windows need to be measured up by Cherie, Glacier Glass' local representative. The measurements will be sent back to the factory for the custom double glazed panels to be manufactured.

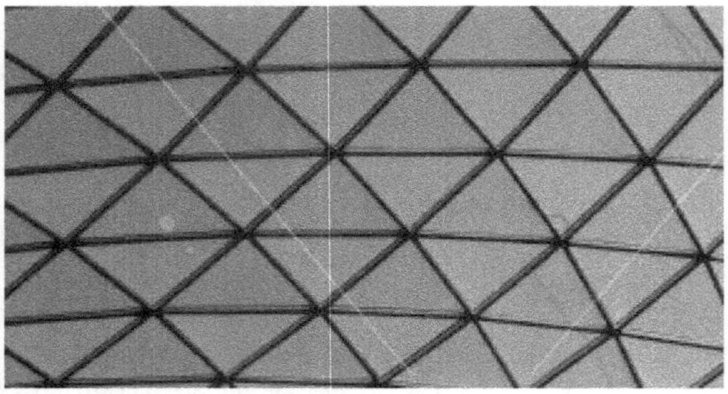

What measurements are sufficient to ensure that the glass panel manufactured will fit perfectly?

When the triangular space for the window and the manufactured window are the same size and shape, we say that the two triangles are congruent.

The activities in this chapter develop the ideas of congruence and similarity. Cherie takes three measurements of the window and the labs explore whether these measurements are sufficient to ensure that the manufactured window will fit. If the triangular panel made to specification **must** fit then all triangles made to those specifications are congruent. That is two triangles with those specifications have the same size and shape and are congruent. The 4 congruent triangle postulates are developed from drawing and making the triangles.

In order for students to appreciate and accept the postulates the labs are designed to follow pencil and paper construction techniques and to give students the opportunity to explore the constructions. It could be a valuable class discussion as to why these techniques are used as is doing pencil and paper constructions. The advantage of using GX is that the drawing can be done quickly enabling a greater focus on the key idea. Which sets of conditions give rise to triangles identical in size and shape?

Summary of the scenario's process:

- Cherie makes three measurements
- These measurements are sent back to the factory.
- The glass to fit the window is manufactured.
- The glass panel is delivered for installation. Will the panel fit?

An introductory lesson

If time permits using a lesson to simulate the scenario would provide a sound basis for developing the concept of congruence.
Materials required:

- Cardboard
- Scissors
- Protractor
- Ruler
- Pre-prepared Triangular windows, all different scalene triangles. Cut these from corrugated cardboard and label each one.

Explain the above scenario and ask each student to take Cherie's role. They measure their *window* using the minimum number of measurements. Their written measurements will then be passed on to another student who will be the manufacturer. Students then manufacture this *glass panel* from card. When these have been completed the *panels* are installed in the *window*. If you are doing it with more than one class the specifications could be swapped between classes.

It is likely that the activity will raise many issues and opportunities for reflection and discussion such as:

- What measures were provided?
- What was the smallest number of conditions that worked?
- Are there errors in measurement – measuring the window or constructing the panel?
- In practice the panel is made smaller. How much smaller is OK?

LAB # 1 *Three sides*

Aim: Investigate whether or not knowing the lengths of three sides is sufficient to ensure that the factory will construct the window to fit.

ALPINE GLASS

Crystal Clear Custom Windows

Alpine Glass specializes in those difficult to make windows.

Aspen Lodge has asked for their premium double glazed panels for their unique architect designed ski lodge. The triangular windows need to be measured up by Cherie, Glacier Glass' local representative. The measurements will be sent back to the factory for the custom double glazed panels to be manufactured.

Cherie, Glacier Glass' local representative, measures the three sides. She measures 4, 6 and 8 feet.

Part 1

Open *Geometry Expressions* and start a new sketch File | Open.

Construct a side of length 4.

From the Tools panel select the Draw | Line Segment tool.
Click, drag and release is one way to draw the line segment.

Choose the select tool.
Click on the line segment.

Click the Constrain | Distance/Length tool
Type in 4 and press Enter.

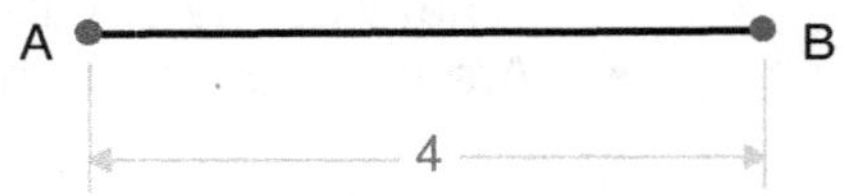

The line segment has endpoints labeled A and B and is of length 4 units.

Draw a circle with center A and radius 6.

From the Draw toolbox select the Circle tool.
Click on A, drag and release.
Select tool
Click on A and then Ctrl-click on C

Click the Constrain | Distance/Length tool
Type in 6.

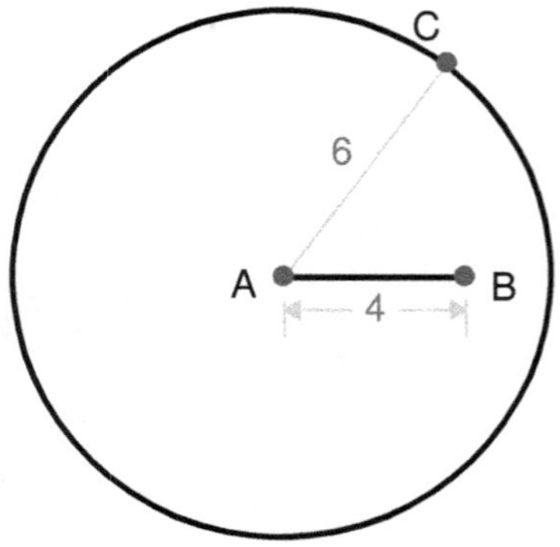

Draw a circle center B and radius 8.

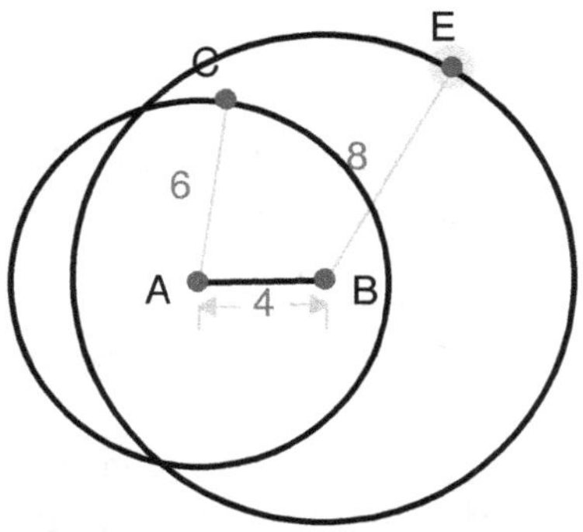

Mark points 6 units from A and 8 units from B.

These are the points of intersection of the circles.

Select both circles by clicking on one and then holding the shift key down click on the other. From the Construct menu select the intersection tool .

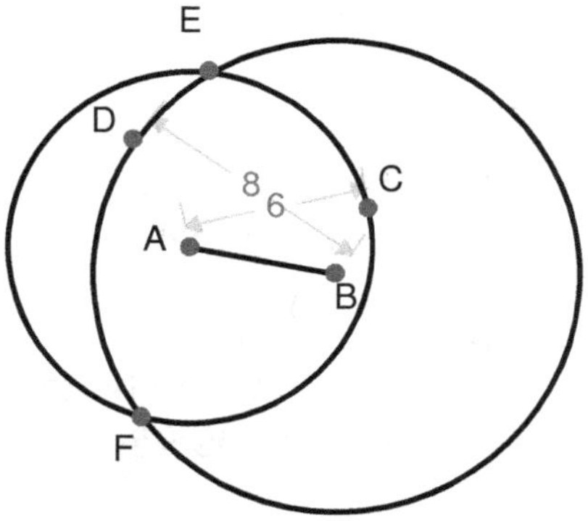

It is also possible to choose the point tool and when the mouse is in the right spot both circles are selected and clicking will mark the intersection.

Draw line segments AE and BE

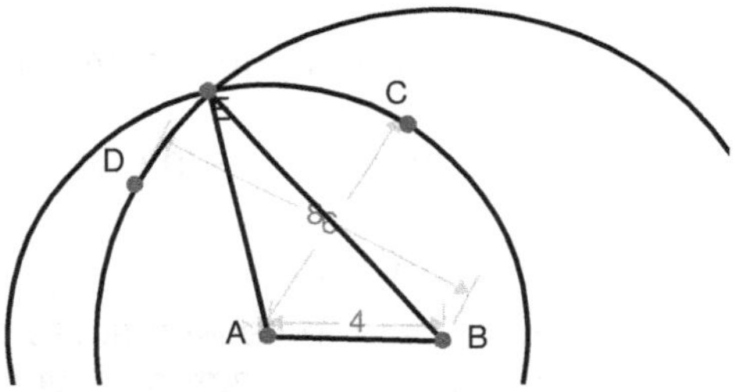

Hide the construction circles

Select tool
Click on circle
Right click and choose Hide
or View | Hide

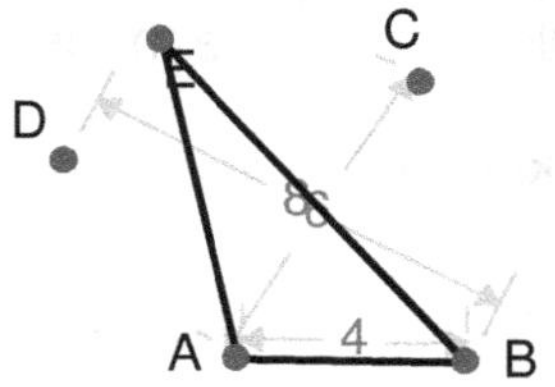

Check that the angle mode is set to degrees, the default setting is radians.

Set the angle mode to degrees
Choose Edit | Settings | Math.
Click the down arrow by Angle Mode to select degrees.

Measure the lengths and angles

Select a side
Choose Calculate (Output) |Real | Length

AB = 4

AE = _____

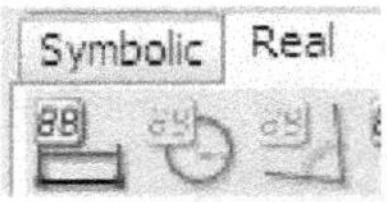

Repeat for each side.
Select two sides (Ctrl-click)

Choose Calculate (Output) | Real | Angle 
Repeat to measure the angles

BE = ____

∠A = _______°

∠B = _______°

∠E = _______°

Drag the vertices of your triangle. Describe what happens. (Think about what changes and what stays the same.)

____

____

Part 2

Open a new sketch

Construct a triangle.
Constrain the lengths to 4, 6 and 8.
Measure the angles.

∠A = _______°

∠B = _______°

∠C = _______°

How do the angles compare to your triangle in Part 1? ________________________

Congruent figures have the same size and shape.

Are the triangles in each sketch congruent? ________________________

Part 3

Build a triangle with sides 5, 12 and 13 units.
Change the lengths on your current sketch.
(Double click on the measurement, and then enter the new value).

What kind of triangle is this? ________________________

Part 4

What happens when you try with sides 4, 5 and 10? ________________________

Why? ____

____

(Note: Press Ctrl-Z to undo the previous step)

If Cherie sends the lengths of the three sides will the manufactured window fit?

____

LAB #1 Three sides

In this activity students:

- Simulate a scale drawing using ruler and compass construction;
- Establish that specifying the three sides ensures the correct window is manufactured (that all constructed triangles are congruent);
- Construct the triangle using *Constraints* and
- Investigate the triangle inequality.

You may like to introduce the context of Glacier glass through a class discussion before beginning the lab activity.

> I.e. Cherie, Glacier Glass' local representative, measures the three sides of the triangular window as 4, 6 and 8 feet. Assuming that Cherie's measurements are accurate and the factory creates a triangular window panel with these measurements accurately, must the window fit? It may be useful to note that in practice the glass would be cut smaller allowing a tolerance for fitting. This would normally be about three eighths of an inch.

The instructions for the GX construction are quite detailed so it can also be used as an introduction to drawing in GX. This includes setting the angle mode to degrees as *GX*'s default mode is radians. For a class that has significant experience of GX you may like to develop a condensed activity sheet with outline instructions only.

A simpler way of creating the 4-6-8 triangle would be to draw the triangle and then constrain the sides. However, students would not have the opportunity to explore why knowing the three sides means that only one triangle can be created.

A further discussion point needing to be addressed is mirror images. It is likely that some students will create mirror images and see these as different triangles as it is not possible to superimpose one upon the other. Flipping the window over will allow it to fit. Having two 4", 6" and 8" cardboard triangles already prepared would be helpful. Draw around the model on the board. Ask a student to measure the "window" and record the measurements. Simulate the mirror image being unloaded on site with the other model and then ask for it to be installed.

Solution

Part 1

AE = 6
BE = 8
$\angle A = 104.5°$
$\angle B = 46.6°$
$\angle E = 29°$
The size and shape stay the same, the orientation and position can change.

Part 2

$\angle A = 104.5°$
$\angle B = 46.6°$
$\angle C = 29°$
The angles are the same.
The triangles are congruent.

Part 3

Right triangle

Part 4

It disappears into a line of length 10.

This is because you can not have a triangle where one side is longer than the sum of the other two sides. This is called the triangle inequality.

If Cherie sends the lengths of the three sides the manufactured window **will** fit?

LAB # 2 *Two Sides and an Angle*

Aim: Investigate whether or not knowing the lengths of two sides and one angle is sufficient to ensure that the factory will construct the window to fit.

Cherie measures one side at 12', a second side at 8' and a 45° angle.

Part 1

Draw a triangle with sides 12, 8 and an angle of 45° between the sides.

Check that the angle mode is set to degrees, Edit | Preferences | Math

Draw a triangle with sides 12, 8 and an angle of 45° between the sides.

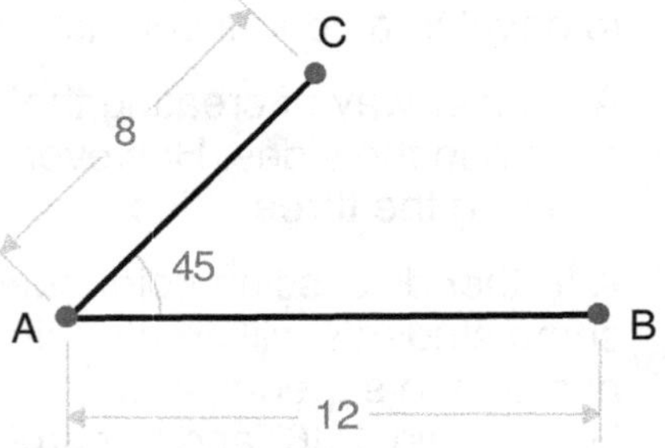

Draw line segment AB
Draw line segment AC Constrain length of AB to 12
Constrain length of AC to 8
Constrain angle BAC to 45°
  Select Lines AB and AC

  Constrain | Angle 
  Enter 45

Draw line segment BC to complete the triangle.
Drag a vertex of the triangle.

Can you change the size or shape of the triangle? ______________________

List the steps (in order) needed to set out the triangle in the factory (using ruler, compass and protractor.

____

____

____

Investigate what happens when you change the angle or side lengths.
Is it always possible to construct a triangle from two sides and the angle between them (SAS)?

____

Part 2

Draw a triangle with sides 12, 8 and an angle of 35° opposite the side of length 8.

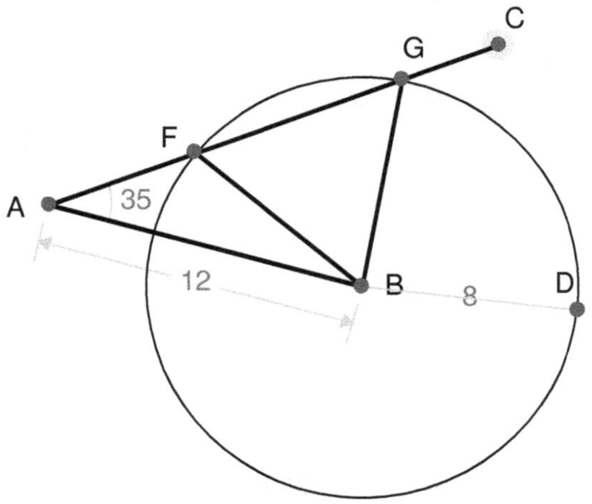

Draw line segment AB

Draw line segment AC

Constrain | Length of AB to 12

Constrain | Angle BAC to 35°

Draw a circle center B
    Make it large enough to intersect line AC twice

Constrain | Radius to 8 

Complete the triangle.
    Make sure both line AC and the circle are highlighted before clicking to begin the line segment

List the steps required to construct these triangles on paper (or a sheet of glass) using drawing tools like compass, ruler and protractor.

____

____

____

Does this set of conditions result in triangles of the same size and shape every time?____

Explain. ____________________________________

____

Vary the conditions by removing the constraint on angle BAC. Drag point C around so that :

Just **one** triangle is possible.

Draw the sketch in the box,

No triangle is possible.

Draw the sketch in the box.

Will specifying two sides and an angle ensure that the correct window will be made? ___

Summarize your findings from this activity. ____

____

____

LAB #2 Two Sides and an Angle

In this activity students:

- Construct a triangle from two sides and the included angle;
- Construct a triangle from two sides and a non-included angle and
- Investigate the ambiguous case.

If students have not recently completed Lab 1 then it will be necessary to outline the scenario. See Lab 1. Cherie measures a client's triangular window on site and sends the measurements back to the factory for the window to be manufactured. It may also be necessary to begin by ensuring that the angles are going to be measured in degrees. Instructions are in Lab 1 [Edit | Settings | Math and then set Angle mode to degrees from the pull down menu].

In Part 1, students investigate the SAS postulate. It should be clear from the construction process that the measurements ensure a unique triangle is drawn. It will be valuable to draw this out in students' reflections on the Lab activity.

In Part 2 the angle is not between the given sides. This leads to the ambiguous case. Drawing the triangle follows a pencil and paper method using a compass, protractor and ruler. Give students time to explore the possibilities. In summarizing you may like to point out that the case for only one triangle occurs when the line is tangent to the circle. This leads into the R-H postulate for Lab 3.

An extension would be to ask students to articulate the *Constraints* on the angle for each of the cases: *i.e.* two triangles (the ambiguous case), one triangle and no triangles.

Solution

Part 1

No the size and shape of the triangle stay the same.

- Measure a side with a ruler or tape.
- Measure the angle with a protractor.
- Measure the length of the second side using a tape along the second ray of the angle.
- Draw in the third side.

Yes. It is always possible to construct a triangle from two sides and the angle between them (SAS).

Part 2

- Measure a side with a ruler or tape.
- Measure the angle with a protractor.
- Set a compass to the second length and swing an arc to intersect the second ray of the angle.
- Draw in the third side. (note there is likely to be two possible triangles.)

No. This set of conditions doesn't necessarily result in triangles of the same size and shape as we have just drawn two differently shaped triangles from the given information.

Part 3

Just **one** triangle.

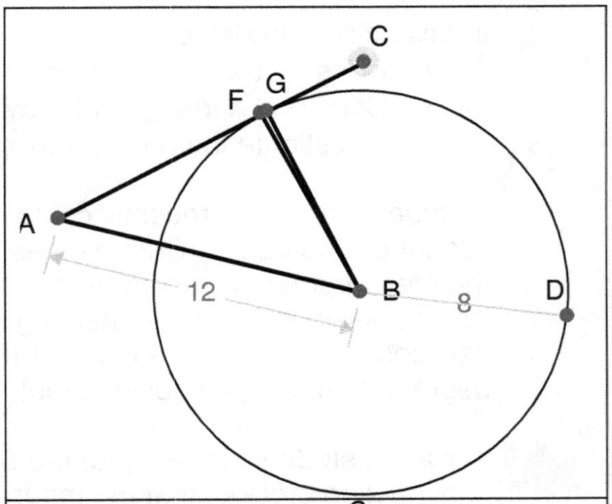

The points F and G become coincident. It may not possible to achieve this exactly.

At this point ∠AFB is 90°and AC is tangent to the circle.

No triangle.

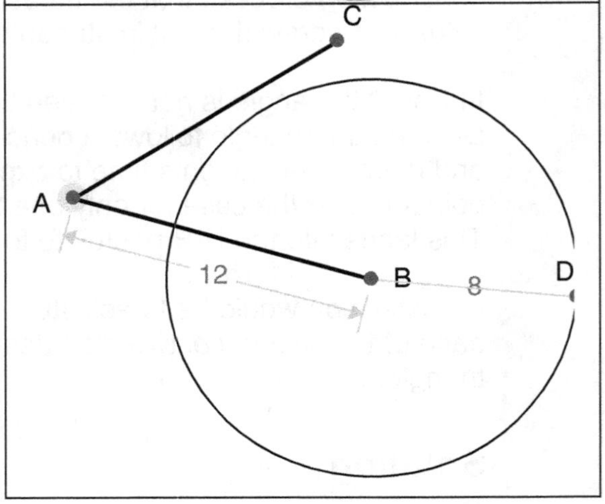

As the circle does not intersect the second ray of the angle no triangle is possible..

No. Specifying two sides and an angle will not ensure that the correct window will be made.

Two sides and the included angle, SAS, means only one triangle is possible.

Two sides and a not included angle is unclear. There may be two, one or no triangles possible with the specified measurements.

LAB # 3 *Hypotenuse and a Side*

Aim: Investigate whether or not knowing the triangle is right-angled, the length of the hypotenuse and another side is sufficient to ensure that the factory will construct the window to fit.

Cherie measures a right angle, the hypotenuse is 12' long and another side is 8' long.

This is a special case of two sides and an angle.
Construct a right triangle with hypotenuse 12 and one leg 8.

Construct a triangle (3 line segments)

Constrain one angle to 90°

Constrain the hypotenuse to 12

Construct a circle from one end of the hypotenuse.

Constrain the radius to 8.

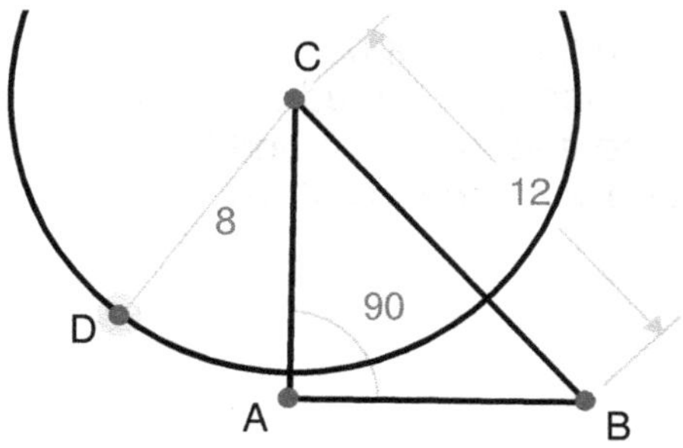

Drag the right angle vertex until it lies on the circle. In the diagram the point A will be dragged.

It is possible to use the Construct | Intersection tool.

Select the circle

Select the vertex of the right angle

Click 

Is it possible for more than one triangle of this size and shape to be drawn from this information?

____

Calculate the length of the third side (using the Pythagorean Theorem)

____

____

____

Is this the same as SSS (Lab #1) ? ______________________

LAB #3 Hypotenuse and Side

In this activity students:

- Construct a right triangle with given hypotenuse and side;
- Establish that specifying the hypotenuse and side of a right triangle ensures the correct window is manufactured (that all constructed triangles are congruent) and
- Use the Pythagorean theorem to see this a special case of the SSS postulate

This is a short lab and can be seen as a special case of the exploration of two sides and an angle in Lab 2 part 2.

In the diagram opposite, if ∠AGB is a right angle then BG is a radius of the circle, AG a tangent to the circle and there is just one possible triangle AGB.

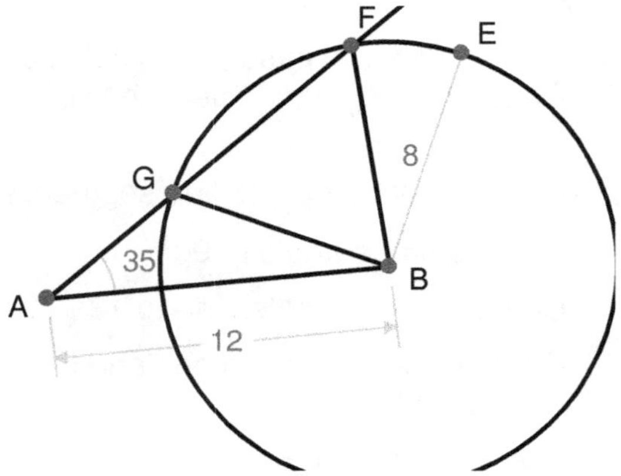

An alternative justification uses the Pythagorean Theorem. Given the hypotenuse and one leg of a right triangle use the Pythagorean theorem to calculate the third side, *i.e.* three sides are known, thus satisfying the SSS postulate (Lab 1)

Solution

No. Only one triangle of this size and shape can be drawn from this information.

$$12^2 = 8^2 + x^2$$

$$x^2 = 144 - 64 = 80$$

$$x = \sqrt{80} \approx 8.94$$

Knowing two sides of a right triangle means the third side can be calculated *i.e.* it is specified. This is the same as knowing three sides.

LAB # 4 *Two Angles and a Side*

Aim: Investigate whether or not knowing the lengths of one side and two angles is sufficient to ensure that the factory will construct the window to fit.

Cherie measures one side at 4', an angle of 38° and a second angle of 75°.

How many different triangles with angles of 38° and 75° with a side of length 4 units can be drawn? Draw as many as you can.

Draw a triangle with one side 4 units long.
  Draw three line segments
  Constrain one side to length 4.

Set two angles to 38° and 75°
  Select two sides (Ctrl-click) and constrain the angle.
  Repeat for a second angle.
  (The diagram shows one possibility with the 4 opposite the 38°)

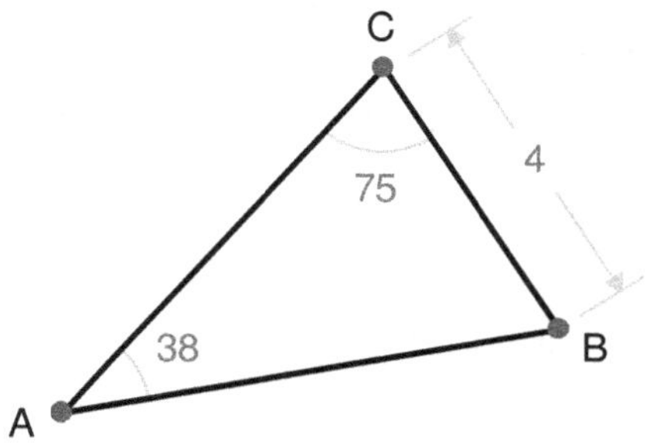

Draw a triangle of different size and/or shape from the given information.
(Hint: place the length 4 opposite a different angle)

Make a drawing in the box.

Draw a third triangle with different shape.

Make a drawing in the box.

What information is needed to determine which triangle Cherie is referring to?

____

____

LAB #4 Two angles and a Side

In this activity students:

- Construct a triangle from one side and two angles and
- Establish that it is necessary to know the position of the side relative to the angles for congruence.

The positioning of the side in relation to the angles is crucial to understanding the AAS or ASA postulate. In this activity students are asked to explore and generate the three possible triangles created by positioning the known side opposite the remaining angles.

While summarizing the results point out that when two angles are known then the third angle can be calculated. Two triangles will be congruent when the known side is opposite the same angle and a second angle is also congruent.

Solution

Three triangles are possible

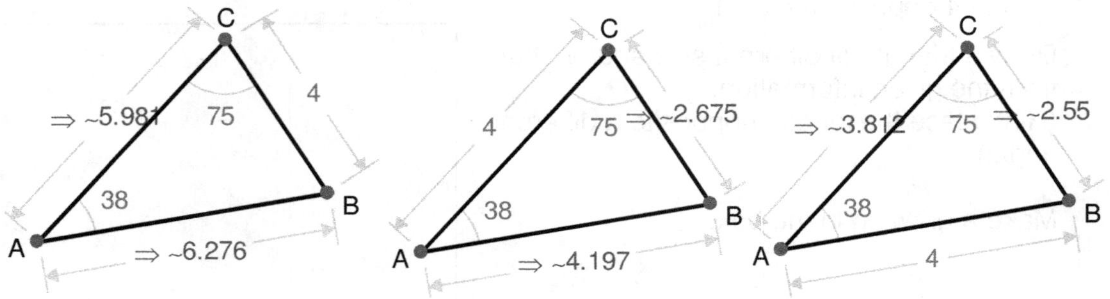

The position of the side of length 4 needs to be known.

Specifying which angle the 4 is opposite will determine the triangle. Sometimes this congruence condition s referred to as ASA meaning that the known side is between the angles. Note that if two angles are known then the third can be calculated.

LAB # 5 *Three Angles*

Aim: Investigate whether or not knowing three angles in a triangle is sufficient to ensure that the factory will construct the window to fit.

Construct a triangle with angles of 53°, 82°.

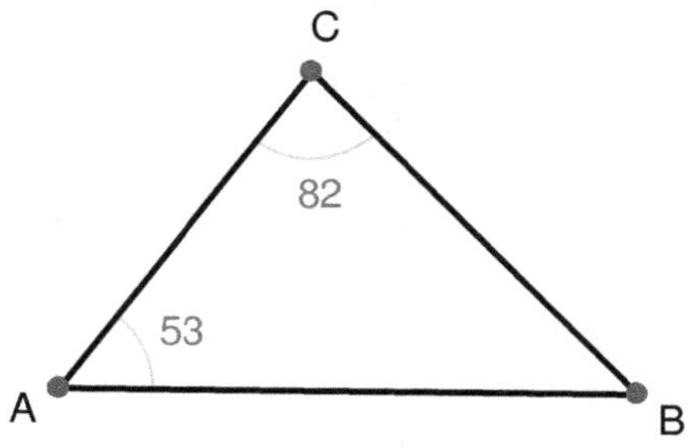

What is the size of the third angle? ____________

Construct three line segments
Constrain the angles

Change the Settings to facilitate the calculations:
Edit | Settings | Math and make Show Name - True.

Output	
Use Assumptions	False
Use Intermediate Variables	False
Show Intermediate Variables	True
Show Name	True

You may change the number of decimal digits displayed. 2 places make the values easy to read.

Math	
Angle Mode	Degrees
Intermediate Variable Complexity (2 to 100)	15
Decimal Digits (0 to 8)	2 Digits

Now Calculate the lengths of the sides.

Symbolic | Real

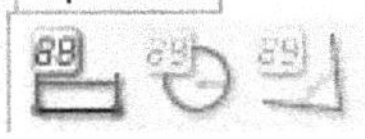
From the calculate menu, click the Real tab and choose length .

Drag the vertices around.
What changes about the triangles and what stays the same?

____

____

Triangles with the same shape are called similar. They do not need to be the same size.

Use the side's name (z_x) to calculate the ratios of the sides.

Create an expression

Use the symbols menu or square brackets to indicate subscripts: z[0] / z[2] . Enter the expressions show to the right.

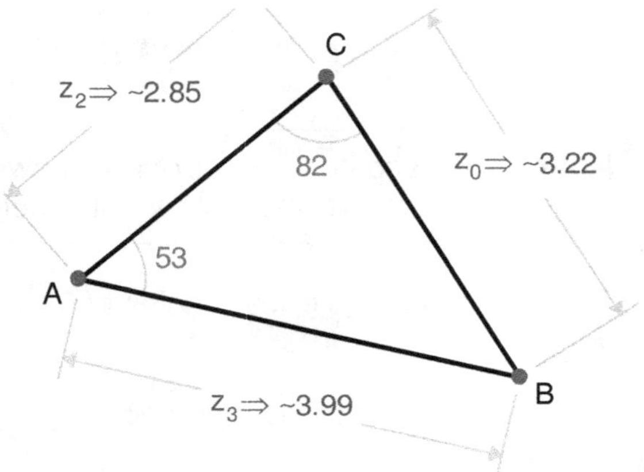

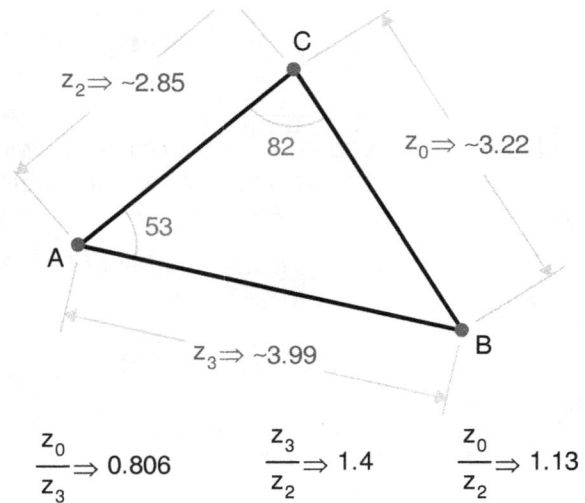

$\frac{z_0}{z_3} \Rightarrow 0.806$ $\frac{z_3}{z_2} \Rightarrow 1.4$ $\frac{z_0}{z_2} \Rightarrow 1.13$

What do you notice about the ratios as you drag the vertices around?

____

____

Chapter Summary:

What sets of conditions produce congruent triangles? (A triangle of the same size and shape)

____

____

____

____

List a set of conditions which will not necessarily produce identical or congruent triangles.

____

LAB #5 Three Angles

Aim: Establish that three angles is not a sufficient condition for triangles to be congruent but is sufficient for similarity.

In this activity students:

- Construct a triangle given three angles;
- Observe that the size is not specified;
- Calculate the ratio of corresponding sides and
- Observe the ratios are equal

Previous labs in this chapter have looked at three facts about a triangle. This lab completes the exploration by exploring the case for three angles.

In Part 1, students create a triangle with specified angles. Dragging vertices of the figure enables them to observe how the size changes and the shape stays the same.

In Part 2, students use GX to calculate the ratios of the sides. As the vertices are dragged around they can observe that the ratios remain constant.

To explore the minimum conditions required to establish similarity would make an excellent Extension activity.

Solution

45°
The shape remains the same. The size, orientation and position can all be changed. The three ratios remain unchanged.

The triangles are called similar triangles and the ratio of corresponding sides being in the same proportion is a sufficient condition for similarity.

Chapter 6 – Proof

Teacher Notes

Proof is about explaining why. Many students find proof in geometry tedious and do not understand the process or the results. Dynamic geometry software can assist students to discover and/or develop understanding of geometric relationships.

This chapter builds from the chapter on congruence. The activities begin with traditional compass and straight edge constructions. These methods are explained using GX's tools, involving students in carrying out those constructions in the computer environment. Many students will benefit from also carrying out the constructions on paper with straight edge and compass as such activities support a kinesthetic learning style and variety to classroom experiences. A number of the labs specifically direct students to perform the construction. Where it is not done you may choose to include this activity as and extension.

Once students understand the construction they can then go about proving that the construction method works. Most of the proofs involve congruent triangles as an intermediate result.

The proof is the answer to the question "Why does it have to work?" The figures from the constructions provide the context.

The constructions include

- bisecting an angle;
- drawing a parallel line through a given point;
- copying an angle
- drawing a perpendicular bisector of a line segment.

Other activities are

- central angle theorem;
- other circle theorems
- centers of a triangle.

Proof is explaining reasoning. Levels of explanations and justifications depend upon the audience. For example a high school proof will be very different to the kind of proof a professional mathematician will provide because of what is assumed to be known by the reader. The high school student is satisfying their teacher that they present a logical argument and in geometry, work on proof often begins with Euclidian definitions and postulates. Consequently, the proofs presented here are unlikely to meet every teacher's specific requirements for their students. The labs can be adjusted appropriately. The emphasis is on presenting the geometrical reasoning.

The activities in this chapter begin with proofs based on using the congruence postulates.

LAB # 6 *Congruent triangles - SSS*

Aim: To write a proof for two triangles being congruent.

Two triangles have the same side lengths as each other.

Step 1: Explore triangles with sides 3, 5 and 6 units.

Open GX and draw two triangles with side lengths 3, 5 and 6.

Draw triangle ABC

Draw triangle DEF

Set the side lengths using the Constrain | Length tool

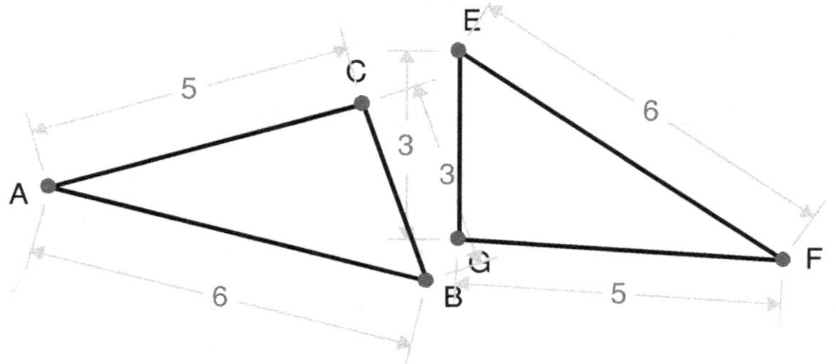

Convince yourself that the triangles are the same size and shape.
Hide the *Constraints*

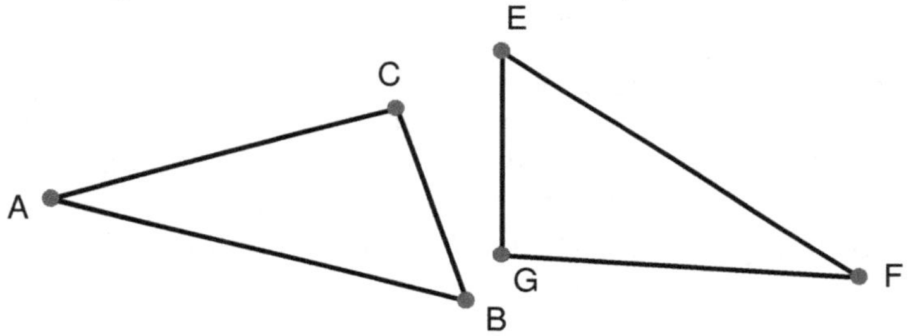

Try dragging vertices around. You are likely to find that the figures move in unexpected ways.

It will be easier to constrain a pair of corresponding vertices (e.g. A and F) to the same coordinates.

Select A, Constrain | Coordinates
Type in 0,0 and press the enter key
Select F, Constrain | Coordinates to (0,0)

Drag the other vertices to make the triangles overlap each other.

Note: mirror images have the same size and shape as well. In some cases you may need to flip one triangle (try the Reflection tool) first to make it match.

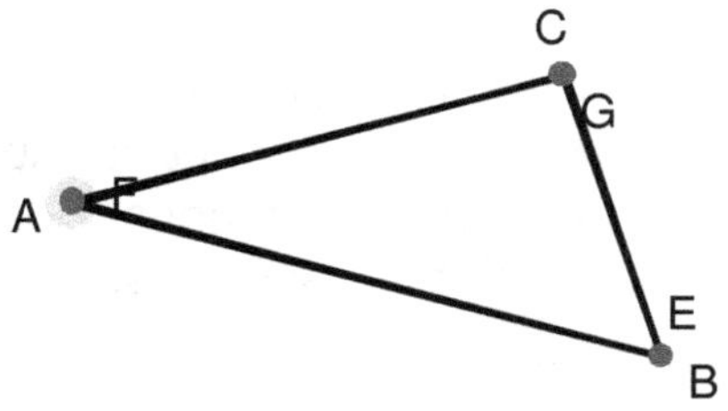

Step 2: Write a proof

Remove the *Constraints*.

Select one side of each triangle and

Constrain | Congruent 

(Note: a dash mark is added to the lines to show they are the same length)

Repeat for each pair of sides.

Again drag to make them match, showing that they are the same size and shape.

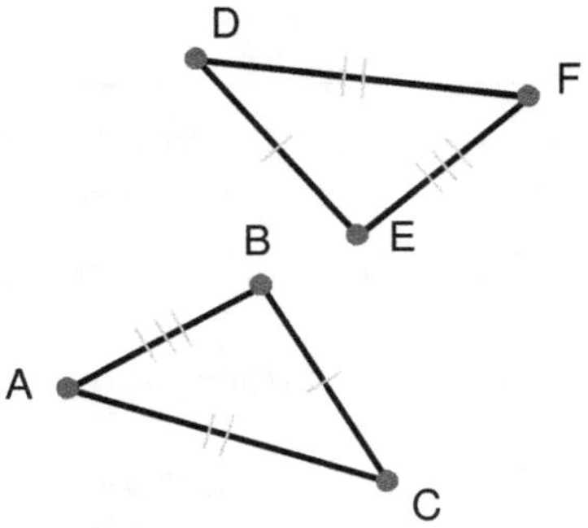

These two triangles are congruent because corresponding sides are of equal length.

A proof may be written in paragraph form.

BC is the same length as DE, AC is the same length as DF and AB is the same length as EF. So the triangles are congruent Side-Side-Side.

Or more formally as a two column proof

Required To Show (RTS) $\Delta ABC \equiv \Delta DEF$

BC = DE	given
AC = DF	given
AB = EF	given
$\Delta ABC \equiv \Delta DEF$	SSS

In the congruence chapter the four conditions for congruent triangles were established. The argument was based around what were the minimum information required to specify the size and shape of a triangle. In formal geometry these sets of conditions are accepted as reasonable and are known as postulates.

Two triangles are congruent when one of the following is true:

- Three pairs of equal sides (SSS)
- Two pairs of equal sides and equal angles between the sides (SAS)
- One pair of equal sides and two pairs of corresponding angles equal (AAS).
- Right triangles with equal hypotenuses and one leg (HS).

Proof

Step 3: Write your own proof

RTS: $\Delta ABC \equiv \Delta FDE$

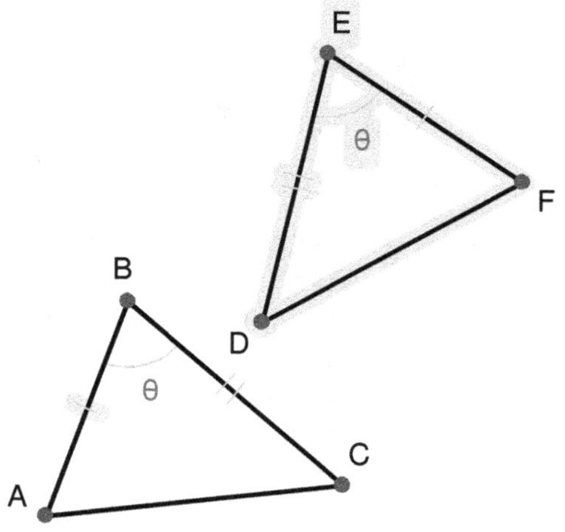

LAB #6 Congruent triangles

In this activity students:

- Draw two triangles with equal side lengths;
- Establish triangles are the same size and shape by dragging;
- Work through a paragraph proof;
- Work through a 2 column proof and
- Write their proof for congruence using SSA.

Some students may create a mirror image for the second triangle and so be unable to drag it on top of the first triangle. They may be quite happy that it is the same size and shape or may need to use a reflection to accomplish the mapping.

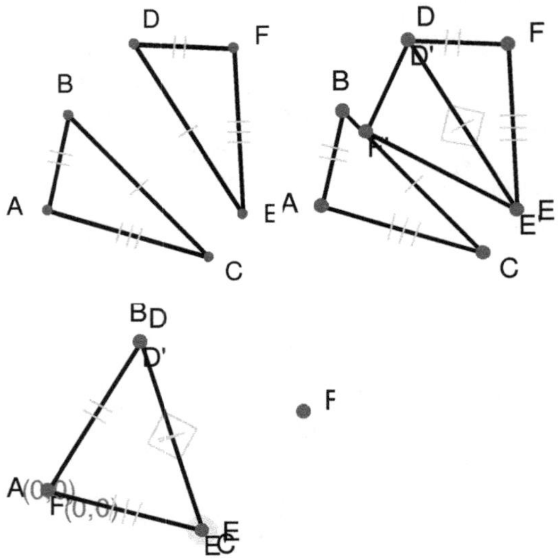

Select the three sides of the second triangle

Construct | Reflection or 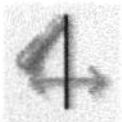
Click on one side of the triangle. In the example to the right we reflect about side DE.

Select the other two sides of the original triangle, right click and choose Hide.

Now the mapping can be completed with the reflected triangle overlapping the original.

After establishing congruence visually using GX, a paragraph and 2-column proof are demonstrated. Students do there own proof for congruence using SAS to complete the activity. You may like to require students to use your preferred method of proof or allow them to choose which ever method they are most comfortable with.

Solution

Step 3:

Required To Show (RTS) $\Delta ABC \equiv \Delta FDE$

$AB = EF$	given
$\angle ABC = \angle DEF$	given
$BC = DE$	given
$\Delta ABC \equiv \Delta FDE$	SAS

LAB # 7 *Bisect an angle*

Aim: Prove the construction for bisecting an angle.

Step1: Understand the construction

Construct an angle

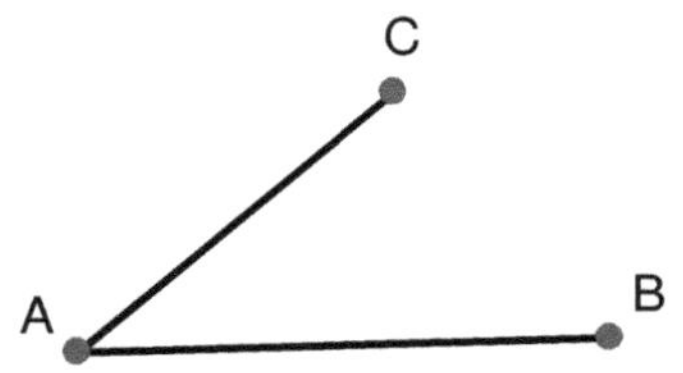

Mark off equal lengths on each ray from the vertex.

Circle tool, drag along AB and release mouse button

Point tool, place the cursor over the intersection of the circle and AC then click.

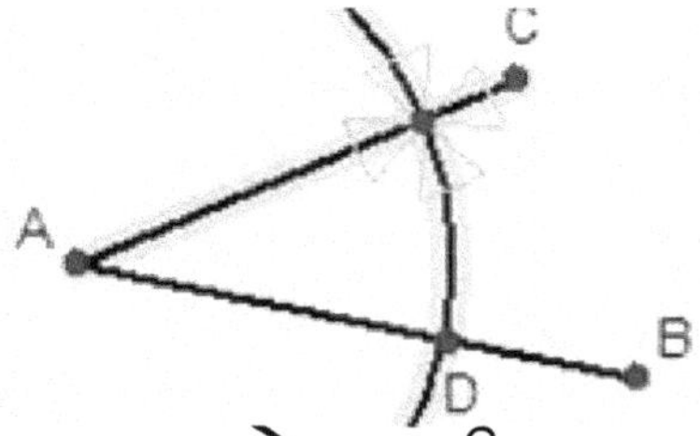

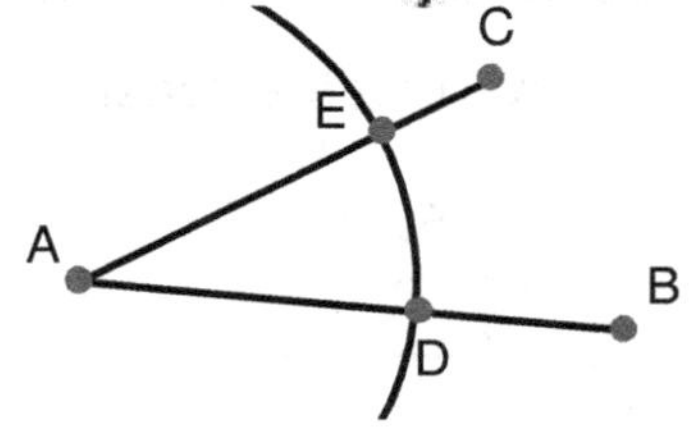

Swing equal arcs from those rays.

Draw circle center D

Constrain radius to be r

Draw circle center E

Constrain radius to the same value

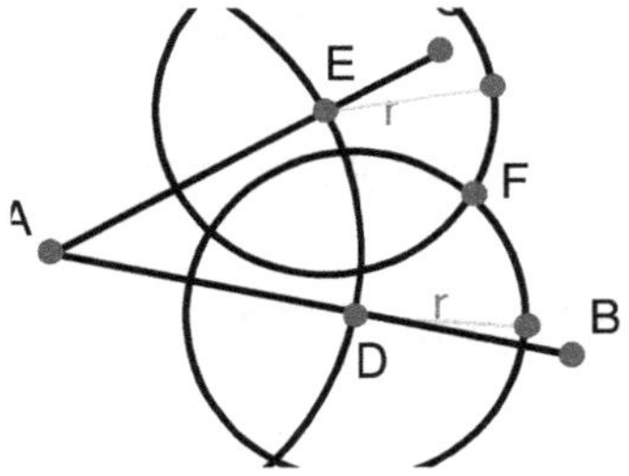

Mark the one of the intersections

Point tool

Draw the line segment AF, the angle bisector

Hide the construction lines (circles).

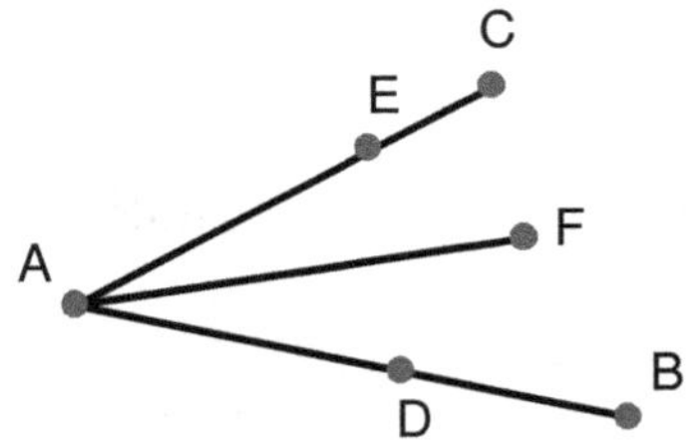

Verify that ∠BAF equals ∠CAF.

Calculate | Real | Angle

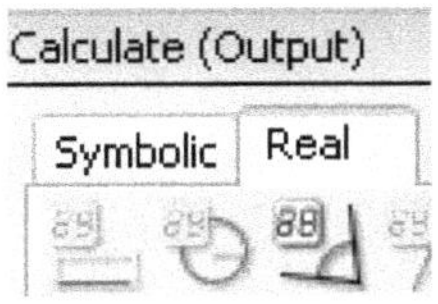

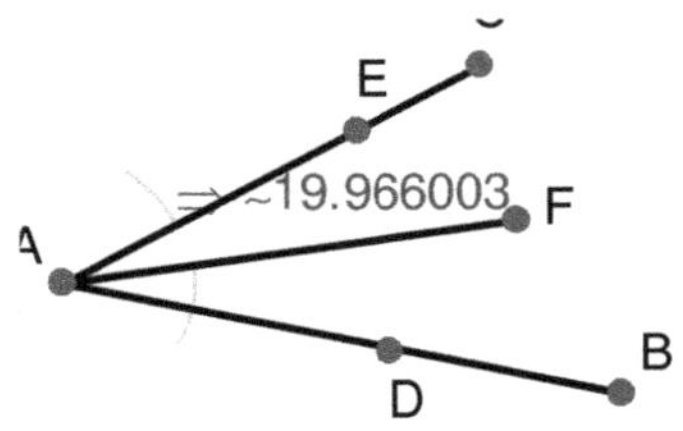

Proof

Drag the points around.

Does AF always bisect ∠BAC? ______________________________

Step 2: Do your own construction using a compass and straight edge.

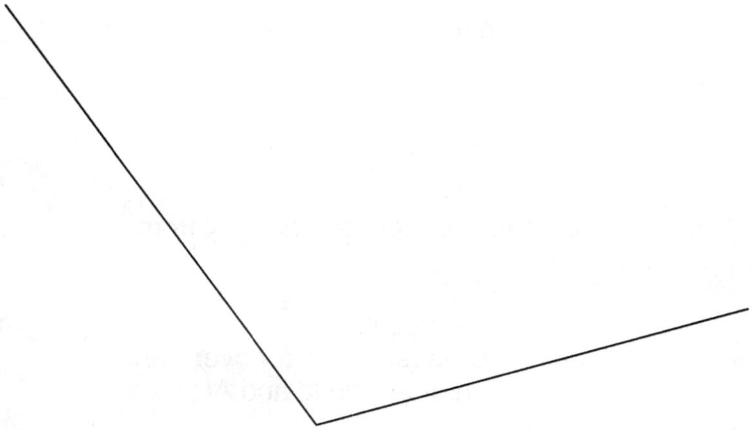

Step 3: Prove that the construction works.

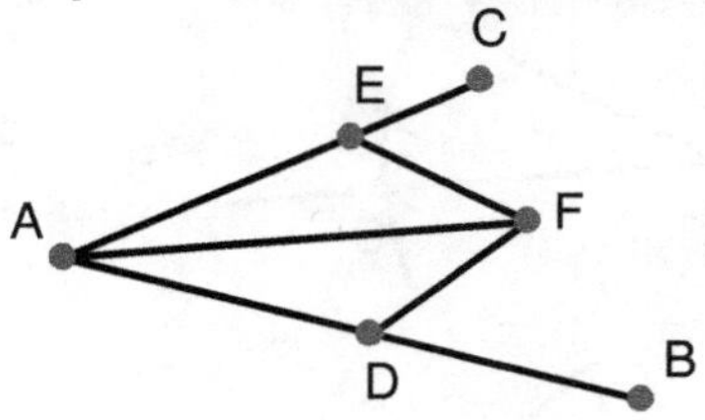

A two-column proof has been laid out below. Provide the missing reasons.

Required To Show ∠DAF = ∠EAF

AD = AE	1. ______________________
EF = DF	2. ______________________
AF = AF	Reflexive property
ΔDAF = ΔEAF	3. ______________________
∠DAF = ∠EAF	4. ______________________

LAB #7 Bisect an angle

In this activity students:

- Simulate constructing an angle bisector;
- Duplicate the straight edge and compass construction on paper and
- Complete a 2 column proof by providing reasons.

The construction to bisect an angle is first carried out using GX. This is quick and flexible in that the angle can be easily changed to see that it remains bisected.

Step 2 is to perform the construction on paper using a compass and straight edge. While this step may be omitted it can support student understanding of the construction method and will be most effective for kinesthetic learners.

Step 3 is the proof. Students have the diagram and are asked to fill in reasons for each step in the proof. Some students may need assistance to see that radii of congruent circles are of equal length; i.e. the use of the compass to mark off equal lengths.

A possible solution is shown below.

Solution

Yes, AF always bisects ∠BAC?

RTS: ∠DAF = ∠EAF

AD = AE	**radii of a circle**
EF = DF	**equal radii**
AF = AF	Reflexive property
ΔDAF = ΔEAF	**SSS**
∠DAF = ∠EAF	**Corresponding angles of congruent triangles.**

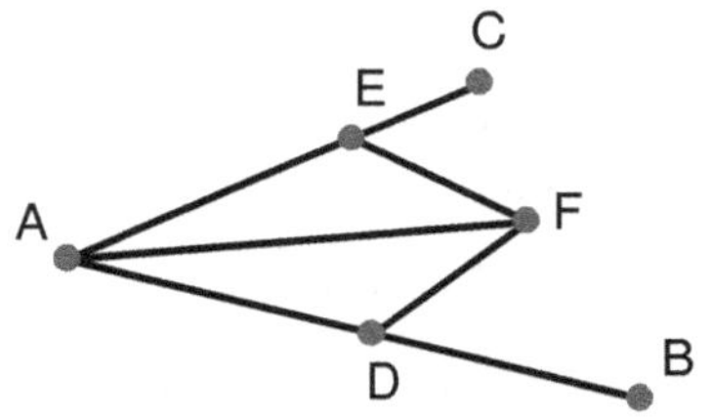

LAB # 8 *Construct a parallel line*

Aim: Prove the construction for a parallel line using compass and straight edge

Part 1: Construct a parallel line.

Draw a line AB
Construct an angle BAC

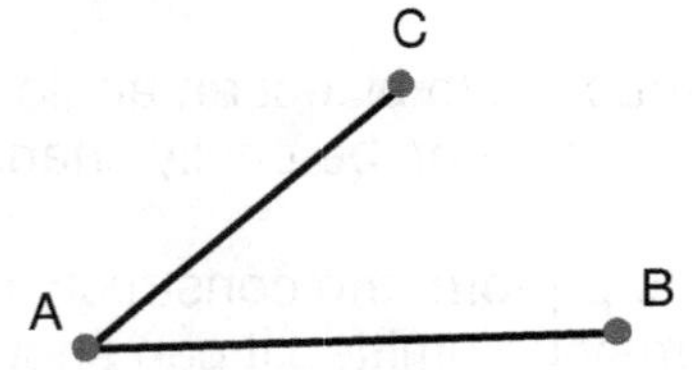

Mark off equal lengths on AB and AC.
- Draw a circle center A
- Use point tool to mark intersections of the circle with AB and AC

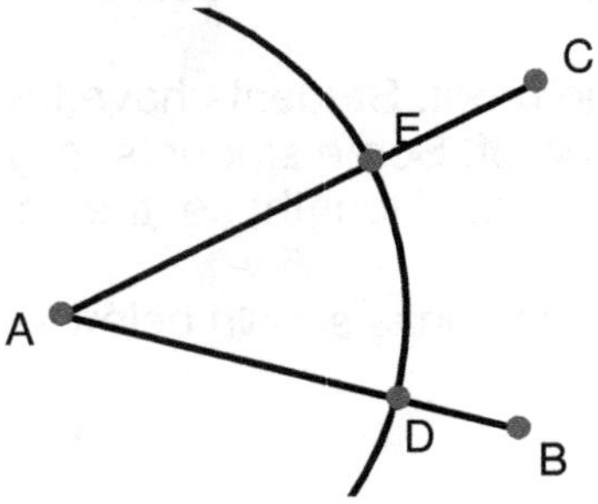

Hide the circle

Draw circle center E
- Drag the circle so it passes through point A

Draw circle center D, also passing through point A

Make the circles congruent
- Select one circle
- Constrain | Radius and set to *r*
- Repeat for the other circle setting radius to *r*.

Mark the intersection of the circles.

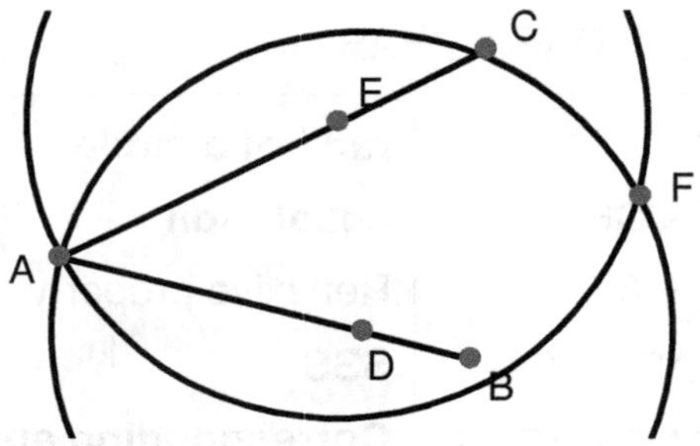

Draw line EF.

Hide the circles

Verify that EF is parallel to AB
Measure ∠CEF and ∠CAB.
- Select lines
- Calculate | Angle

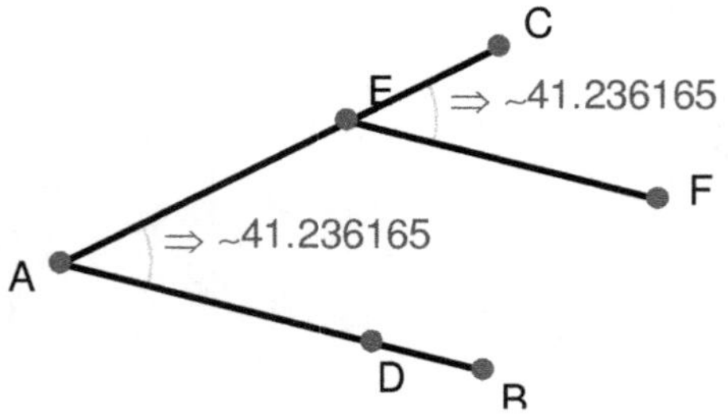

Drag points B, C and D around. Does EF remain parallel to AB at all times? __________

How can you be sure? ____

Part 2: Do your own construction of a parallel line using a compass and straight edge.

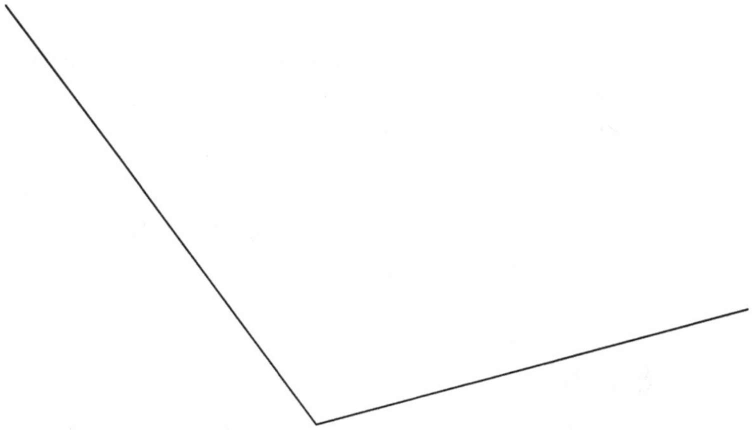

Part 3: Prove that the construction works.

Provide a reason for each statement to complete the proof.

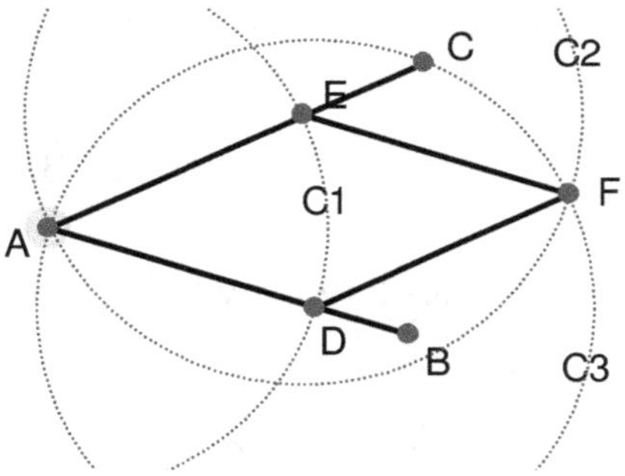

<u>Required to Show (RTS): AD || EF</u>

Let AE = r

EF = r ____________________________

AD = r ____________________________

DF = r ____________________________

ADFE is a rhombus ____________________________

AD || EF ____________________________

LAB #8 Construct a parallel line

In this activity students:

- Simulate drawing a compass and straight edge construction for a parallel line;
- Do the construction on paper and
- Complete the proof

GX has a Construct | Parallel line tool. This is not used in order to emphasize the construction required to develop the proof. The purpose of repeating the construction is to support different types of learners with a range of experiences and to assist in making connections.

Solution

EF || AB where ever B, C and D are dragged.
∠CEF and ∠CAB are corresponding angles. When corresponding angles are equal the lines are parallel.

A possible proof for part 3.

Required to Prove: AD || EF		Notes
Let AE = r		
EF = AE = r	radii of the same circle	*AE, EF - radii of circle C2*
AD = r	radii of the same circle	*AE, AD - radii of circle C1*
DF = r	radii of congruent circles	*AD, DF - radii of circle C3*
ADFE is a rhombus	quadrilateral with 4 equal sides	
AD || EF	opposite sides of a rhombus are parallel	

Notes:

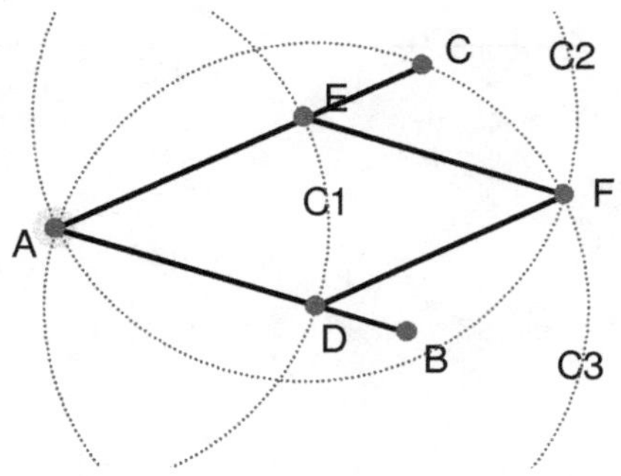

LAB # 9 *Copy an angle*

Aim: Prove the compass and straight edge construction of a congruent angle.

Construct an angle BAC

Mark off equal lengths
Draw circle center A.
Constrain the radius to length *r*.

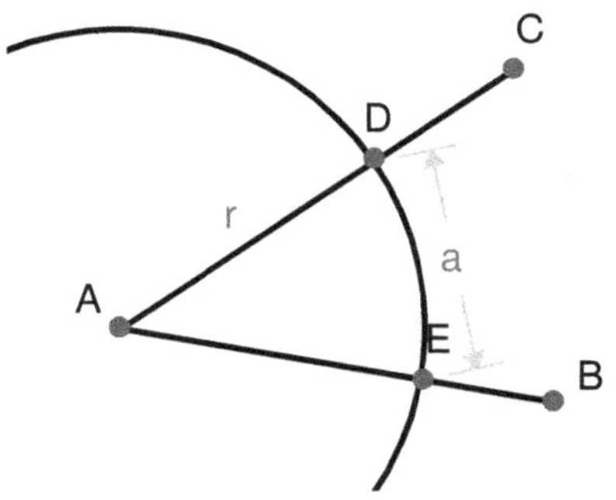

Mark intersection of the circle with AB and AC.

Constrain length DE to default value *a*.

Hide the construction circle
Right click on circle and choose Hide

Draw a second line FG.

Draw circle center F.

Constrain the circle radius to length *r*, (the same as the circle center A).

Put a point I on the circle.

Constrain HI to length *a*.

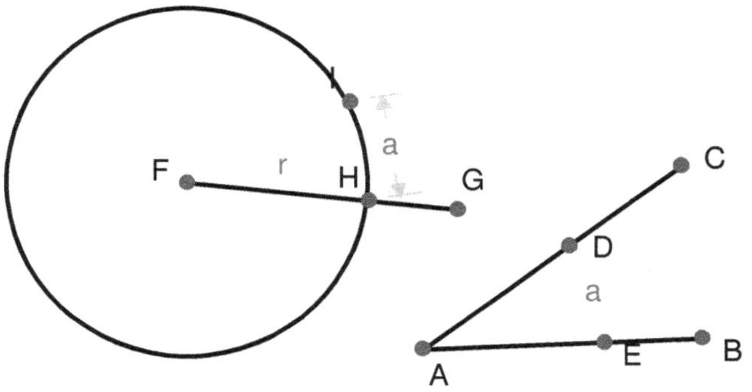

Draw line IF.

Hide the circle.

Measure angles GFI and BAC.

Select each set of lines and
Calculate | Real | Angle

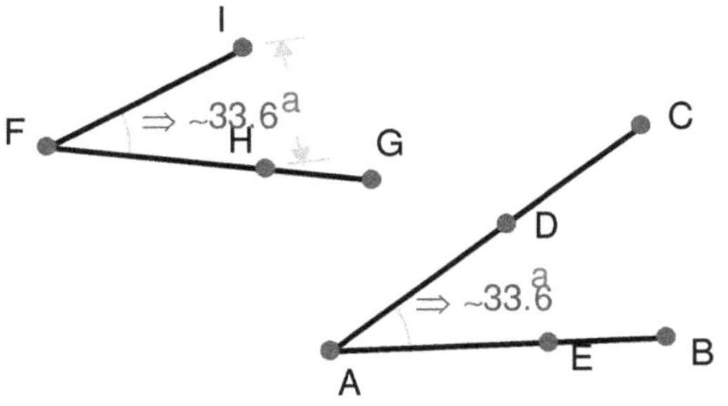

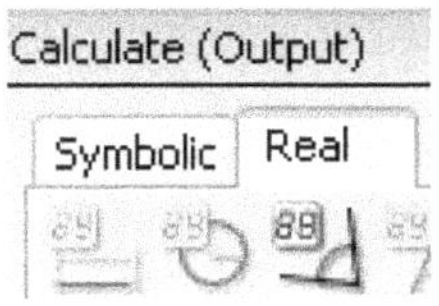

What do you notice about the size of angles GFI and BAC? ____________________

Drag points around and try to find situations where the angles are not equal.

Write a proof for the two angles being equal.
(Hint the angles will be equal if AED and HFI are congruent triangles.)

LAB #9 Copy an angle

In this activity students:

- Simulate the compass and straight edge construction to duplicate an angle and
- Write a proof using congruence triangles.

Duplicating the process using compass and straight edge would be a worthwhile addition to the activity.

No outline is given for the proof for this lab. It is expected that sufficient preparatory work has been done to scaffold student learning and for them to be able to write a complete proof.

Solution

∠GFI =∠BAC?

The angles change but remain equal to each other as points are dragged.

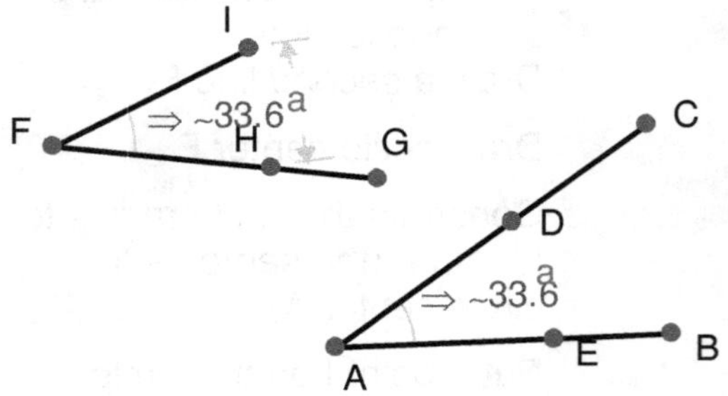

A possible proof

RTS: ∠EAD ≡ ∠HFI

AE = HF	Radii of congruent circles
FI = AD	Radii of congruent circles
HI = ED	Radii of congruent circles
ΔEAD ≡ ΔHFI	SSS
∠EAD ≡ ∠HFI	Corresponding angles in congruent triangles

LAB # 10 *Perpendicular bisector*

Aim: Prove the construction for the perpendicular bisector.

Construct line segment AB.

Draw circles from A and B. Make sure the circles intersect.

Constrain the radius of each circle to the same value, *i.e.* of equal radius.

Locate the two points of intersection
Select both circles and

Construct | Intersection .

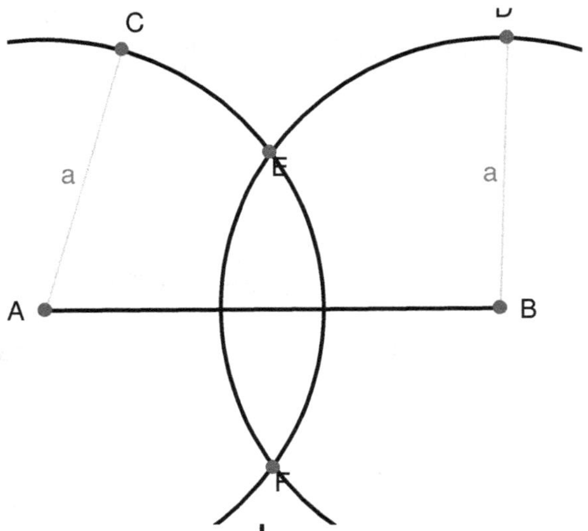

Draw the line connecting the two points of intersection (EF).

Draw point of intersection between EF and AB

Hide the circles and points C and D

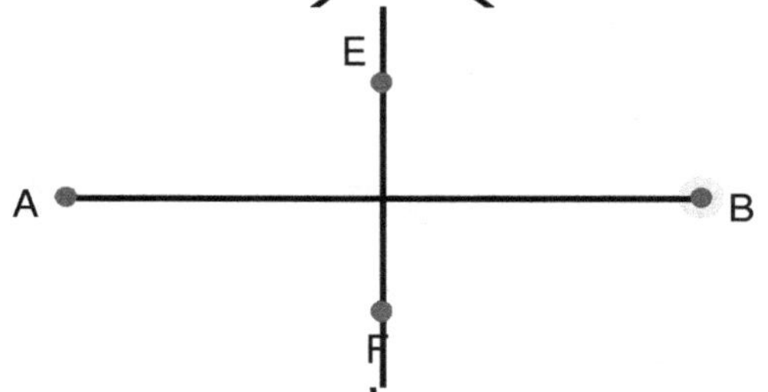

Measure the angle between the lines and the lengths of the line segments.

∠AGE = ________ °

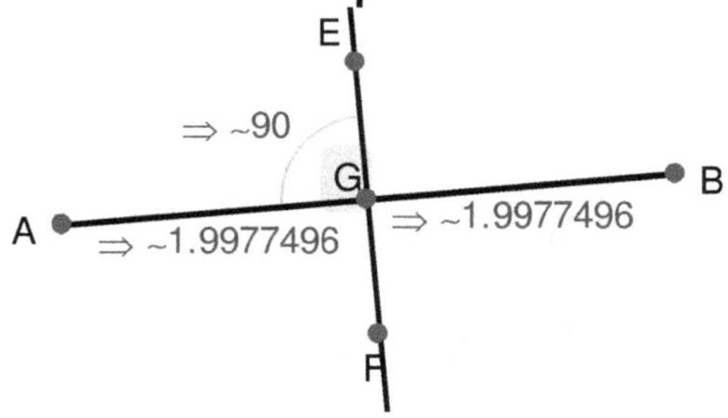

Drag points A and B around.

Does EF always remain perpendicular to AB? ______________________________

What kind of quadrilateral is AEBF? ______________________________

Construct the perpendicular bisector of the line segment using a compass and straight edge.

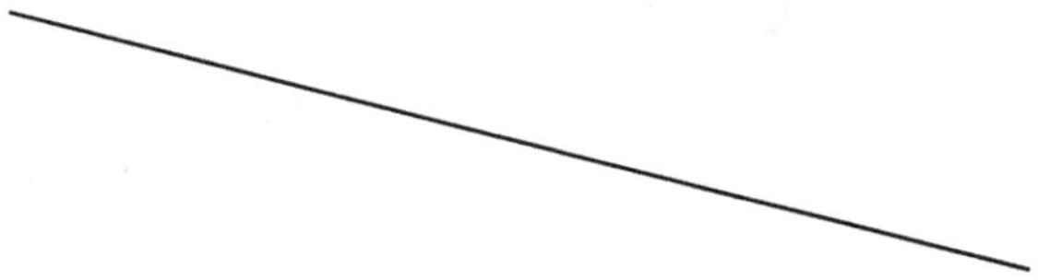

Proof

Why is it so?

AE = AF = BE = BF
radii of congruent circles.

To prove that EF is perpendicular to AB and bisects AB, first show that triangles AEF and BEF are congruent.

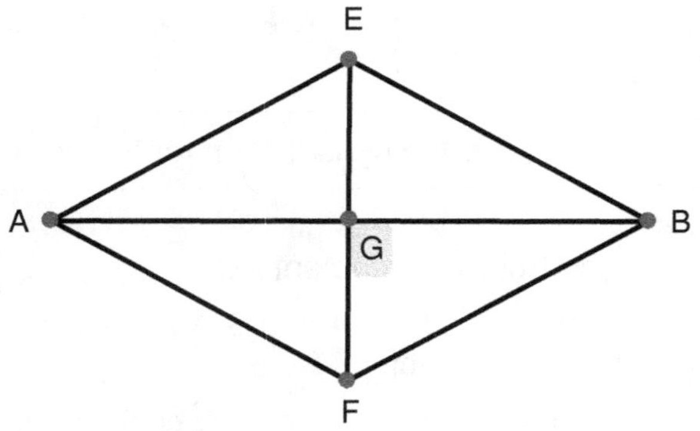

Complete the proof by filling in the blanks. RTS EF bisects AB and EF ⊥ AB

Show ΔAEF ≡ ΔBEF

AE = BE	Radii of congruent circles
AF = ____	
EF = EF	____________
ΔAEF ≡ ΔBEF	SSS

Show ΔAEG ≡ ΔBEG

m∠AEF = m∠BEF	Corresponding angles of congruent triangles
AE = BE	Corresponding sides of congruent triangles
EG = EG	____________
ΔAEG ≡ ΔBEG	____________
AG = BG	Corresponding sides of congruent triangles
That is G bisects AB	
m∠AGE = m∠BGE	____________
m∠AGE + m∠BGE = 180°	Angles on a straight line
m∠AGE = 90°	?
EF is perpendicular to AB	

EXTENSION

Show that the diagonals of a rhombus are perpendicular and bisect each other.
(Hint: AEBF is a rhombus. Show ΔAEG ≡ ΔAFG and EG = FG to complete the proof.)

LAB #10 Perpendicular bisector

Aim: Prove the construction for the perpendicular bisector.

In this activity students:

- Simulate the compass and straight edge construction of the perpendicular bisector of a line segment;
- Perform the construction on paper and
- Complete the proof.

You may wish to omit the compass and straight edge construction depending upon the proficiency and understanding of your students.
The proof is quite involved compared to earlier proofs in this chapter. The steps are presented together with some of the reasons. Students are asked to complete the proof by providing the missing reasons.

Solution

EF is perpendicular to AB.
AEBF is a rhombus.

AE = BE	Radii of congruent circles
AF = BF	Radii of congruent circles
EF = EF	Reflexive property
ΔAEF $\equiv$ ΔBEF	SSS
m$\angle$AEF = m$\angle$BEF	Corresponding angles of congruent triangles
AE = BE	Corresponding sides of congruent triangles
EG = EG	Reflexive property
ΔAEG $\equiv$ ΔBEG	SAS
AG = BG	Corresponding sides of congruent triangles
That is G bisects AB	
m$\angle$AGE = m$\angle$BGE	Corresponding angles of congruent triangles
m$\angle$AGE + m$\angle$BGE = 180°	Angles on a straight line
m$\angle$AGE = 90°	
EF is perpendicular to AB	m$\angle$AGE = 90°

EXTENSION

Show that the diagonals of a rhombus are perpendicular and bisect each other.

(Hint: AEBF is a rhombus.

Show ΔAEG $\equiv$ ΔAFG and EG = FG to complete the proof.)

LAB # 11 *Central Angle Theorem*

Aim: Prove the Central Angle Theorem.

Construct the diagram.
  Construct a circle
  Draw two radii
  Draw chords to a third point on the circle.

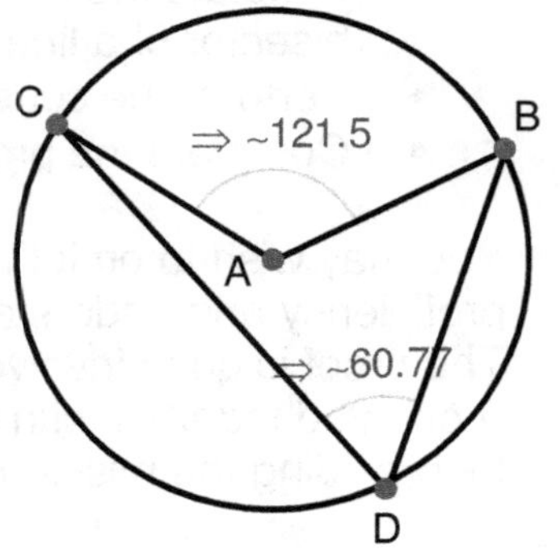

Measure ∠BAC the **central angle**
Measure ∠BDC the angle on arc BC

Drag the figure around.

Make a conjecture involving the central angle. ______________________________

____

Prove that your conjecture is true.
(Hint:

Draw DE passing through A.

∠EAC is the exterior angle to the isosceles ΔACD.

∠EAC = 2 ∠ADC.

Similarly ∠EAB =2 ∠ADB)

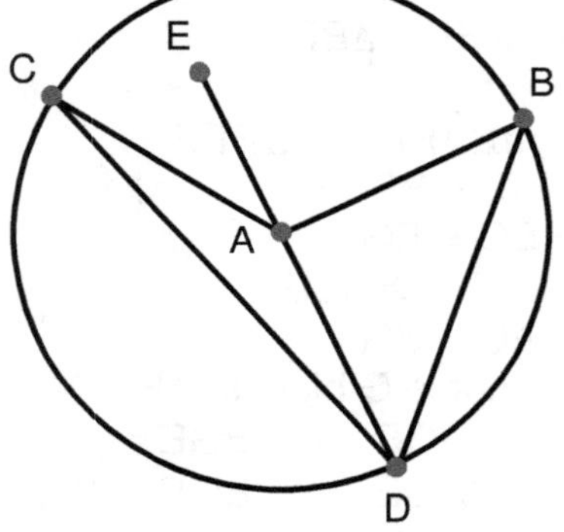

Write a two column proof for the Central Angle theorem.

Required to Show:

LAB #11 The Central Angle Theorem

Aim: Prove the Central Angle Theorem.

In this activity students:

- Construct and measure the central angle;
- Construct and measure the angle subtended by the arc;
- Develop a conjecture and
- Prove the conjecture.

By dragging points around most will observe that the central angle is twice the size. A few students may need prompting; however it is likely they will also be prompted by classmates.

The proof is asked for with no scaffolding. You may wish to provide some guidelines depending upon the skills and experience of your class.

Solution

RTS: $\angle BAC = 2\angle BDC$

Let $\angle ADC = \alpha$ and $\angle ADB = \beta$	
$AC = AD$	Radii of circle
ΔCAD is isosceles	2 sides equal
$\angle ACD = \angle ADC = \alpha$	Isosceles triangle
$\angle EAC = \angle ACD + \angle ADC$	Exterior angle Th.
$= 2\alpha$	
$AB = AD$	Radii of circle
ΔBAD is isosceles	2 sides equal
$\angle ABD = \angle ADB = \beta$	Isosceles triangle
$\angle EAB = \angle ABD + \angle ADB$	Exterior angle Th.
$= 2\beta$	
$\angle BAC = \angle BAE + \angle EAC$	
$= 2\alpha + 2\beta$	
$= 2(\alpha + \beta)$	
$= 2\angle BDC$	

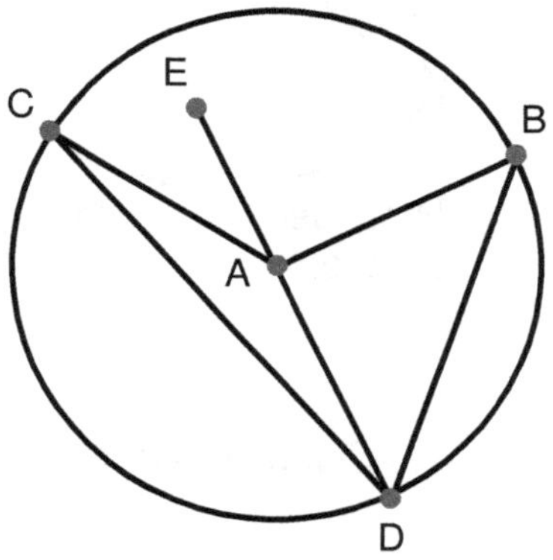

LAB # 12 *Other circle theorems*

Aim: Develop conjectures from diagrams and write proofs.

For each part of this activity construct the diagram in GX.

- Measure angles (and/or lengths) and develop a conjecture from each diagram.
- Drag points in your diagram to see if your conjecture holds.
- Prove (or disprove) your conjecture. (except part 3)

1. Angle in a semicircle. Measure the size of angle D.

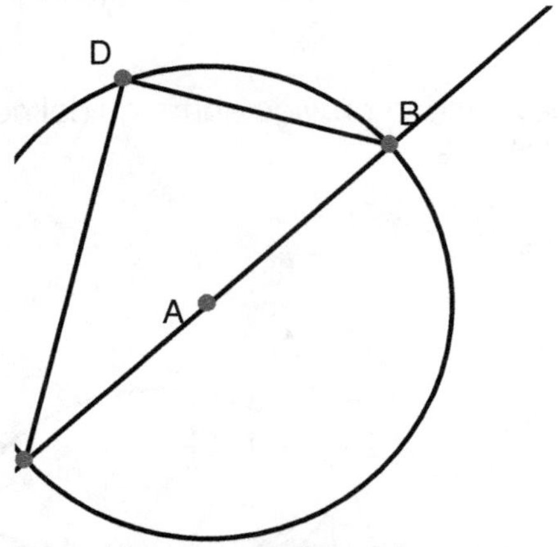

Hint: Use the angle at the centre theorem.

Conjecture:

Proof:

2. Opposite angles of a cyclic quadrilateral. Measure angles B and D.

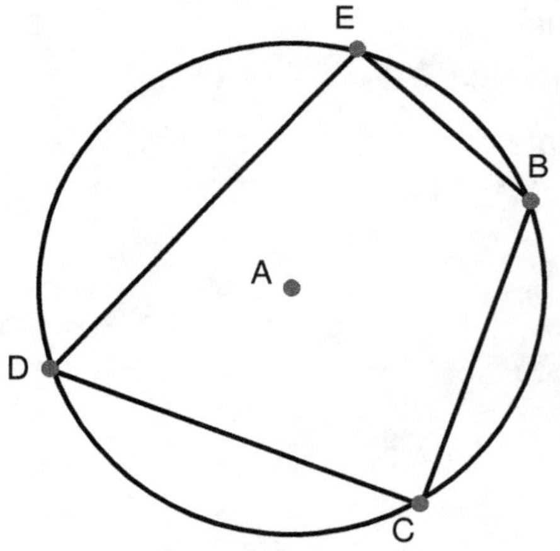

Hint: The angle at the centre theorem can be very helpful.

Conjecture:

Proof:

Proof

3. Tangent and radius. Measure angle between the radius and tangent.

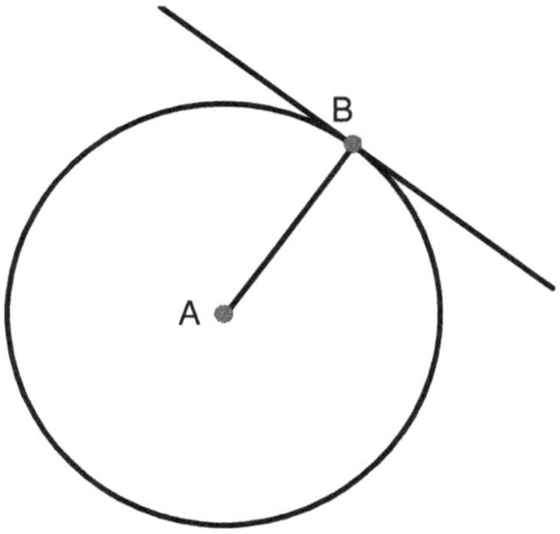

Conjecture:

Constrain | Tangent

4. Tangent and chord. Measure ∠ADF and ∠DEF.

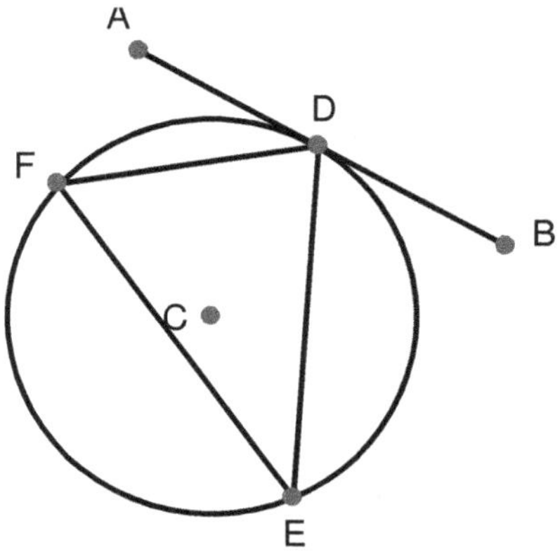

Conjecture:

Proof

5. Tangents from a common point. Measure CG and CF.

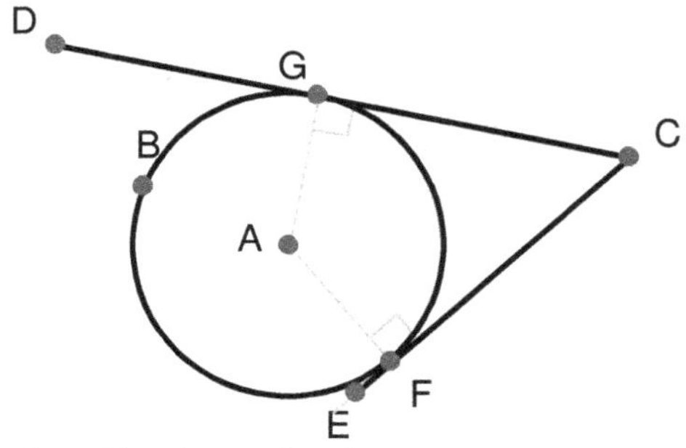

Conjecture:

Proof

Hint: To draw the figure:
- Draw the two lines
- Draw a circle
- Select a line and the circle
- Constrain | Tangent
- Repeat for the second line.

LAB #12 Other circle theorems

In this activity students:

- Construct the diagrams using GX;
- Make measurements (Calculations)
- Develop conjectures and
- Write proofs for conjectures.

This lab requires students to be confident users of GX and to have experience using GX to construct drawings, develop conjectures from patterns and to write proofs. Scaffolding of the proofs may be required.

Solution

1. Conjecture: The angle in a semicircle is a right angle.

Outline for proof
$\angle CAB$ is a straight angle. $\angle CDB$ is half the straight angle because of the angle at the center theorem.

2. Conjecture: Opposite angles of a cyclic quadrilateral sum to 180°.

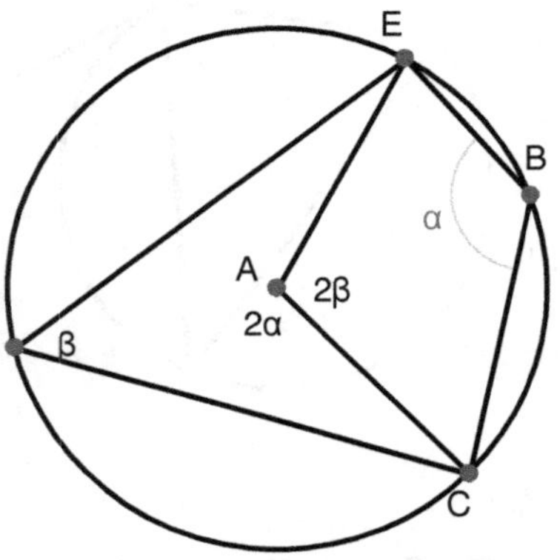

Outline Proof:
Let $\angle EBC = \alpha$ and $\angle EDC = \beta$

$\angle EAC$	$= 2\angle EBC$	Angle at Center Th.
	$= 2\alpha$	
$\angle EAC$	$= 2\angle EDC$	Angle at Center Th.
	$= 2\beta$	
$2\alpha + 2\beta = 360°$		Angle at a point
$\alpha + \beta = 180°$		
$\angle EDC + \angle EBC = 180°$		

3. Conjecture: The angle between a tangent to a circle and the radius is a 90°.

4. Conjecture: ADF = DEF

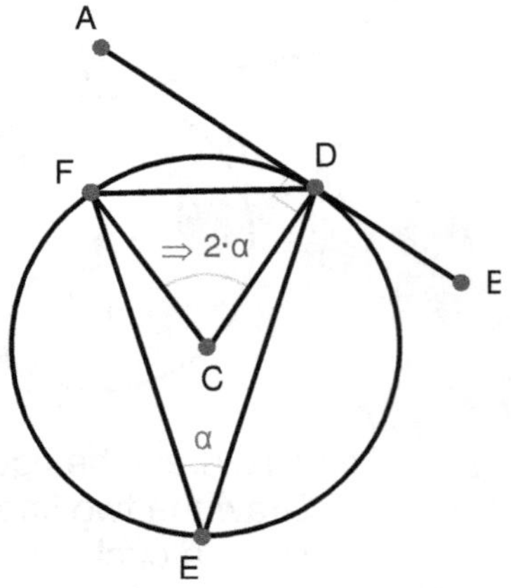

Outline Proof:
Let $\angle DEF = \alpha$
Draw cord FC and CD
$\angle FCD = 2\,\alpha$
$\angle ADC = 90°$
$\angle CFD = \angle CDF$
$= 180° - 2\,\alpha$
$\angle CDF = (180° - 2\,\alpha) / 2$
$= 90° - \alpha$
$\angle ADF + \angle CDF = 90°$
$\angle ADF = 90° - (90° - \alpha) = \alpha$

5. Conjecture: Tangents from a common point are of equal length.

Proof: Establish that ΔAGC and ΔAFC are congruent.

LAB # 13 *The "center" of a triangle*

Aim: To explore "centers" of a triangle.

Where is the center of a triangle? There is more than one answer!

- Follow the directions to construct each of the following drawings within a triangle.
- Explore the properties of each point.
- Match the construction with one of incenter, orthocenter, circumcenter, Steiner point and centroid. Descriptions of these "triangle centers" follow. Write the matching name alongside the construction.

a. Construct the angular bisector of each angle.

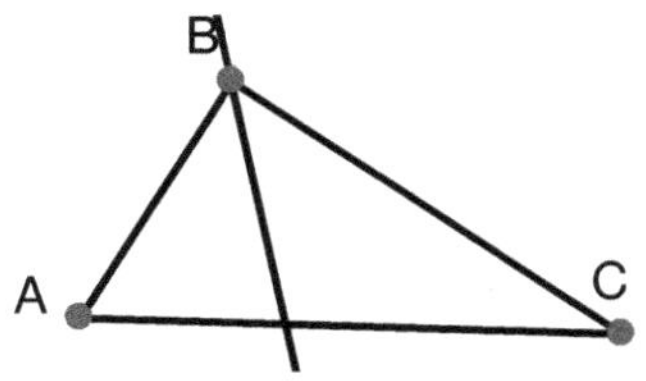

Do they intersect at a common point?

b. Draw the perpendicular bisectors of each side of the triangle.

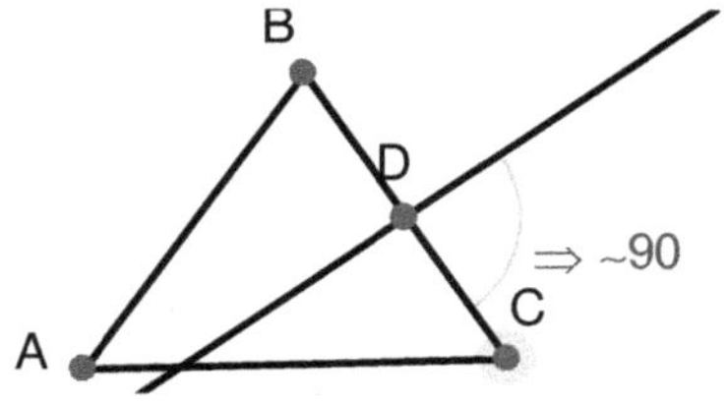

Do they intersect at a common point?

c. Draw lines connecting the midpoints of each side with the opposite vertex.

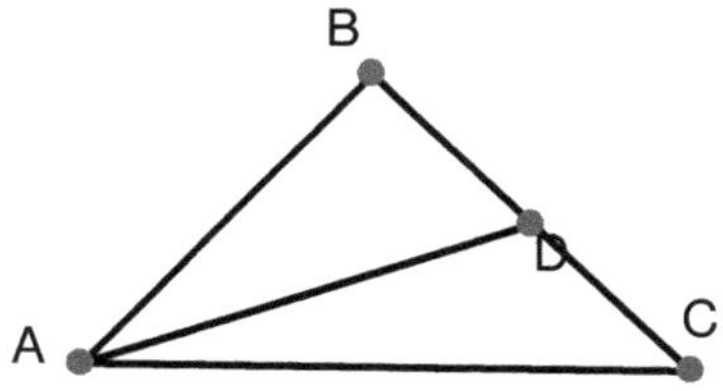

Do these lines meet at a common point?

d. Draw lines through each vertex that is perpendicular to the opposite side.)

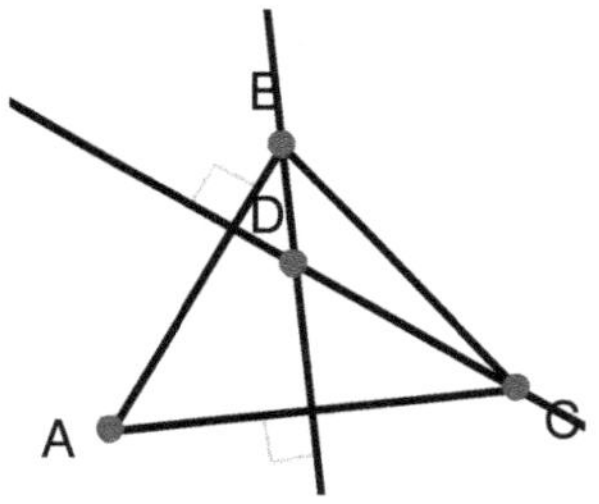

Do these lines meet at a common point?

e. Lines connecting this point to two vertices form an angle of 120°.

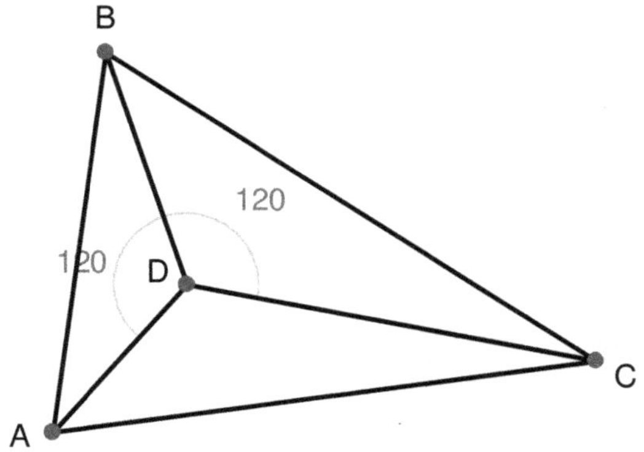

A circle drawn from the **incenter** of a triangle can touch each side of the triangle.

The **orthocenter** of a triangle is formed by the intersection of the altitudes. It does not need to lie inside the triangle.

Proof

The circumcenter is the centre of a circle containing all three vertices of the triangle. It does not need to lie inside the circle.

Sum the distances to the vertices from a point inside a triangle. The minimum value for the sum occurs at the **Steiner point**.

Centroid is the centre of mass, *i.e.* a real triangle can be supported at this point.

EXTENSION

Research proofs for these results.

LAB #13 The "center" of a triangle

In this activity students match a construction with the description of a triangle "center".

The final lab in this chapter briefly introduces five points that could be loosely called centers of a triangle. It is a mix and match activity requiring students to match the diagram with a description of a triangle center. The diagrams include the construction of an incenter, orthocenter, circumcenter, Steiner point and centroid.

This is not an in-depth exploration, rather an introduction to possibilities. It provides a rich source for further Extension activity. Proofs relating to these points are beyond the scope of this chapter.

The aim in this section is to use ***Geometry Expressions*** to develop ideas or conjectures and then provide reasons why our observations must be always true, or if not true when they can be claimed to be true.

Solution

a. Incenter.

b. Circumcenter

c. Centroid

d. Orthocenter

e. Steiner point

Notes:

Incenter

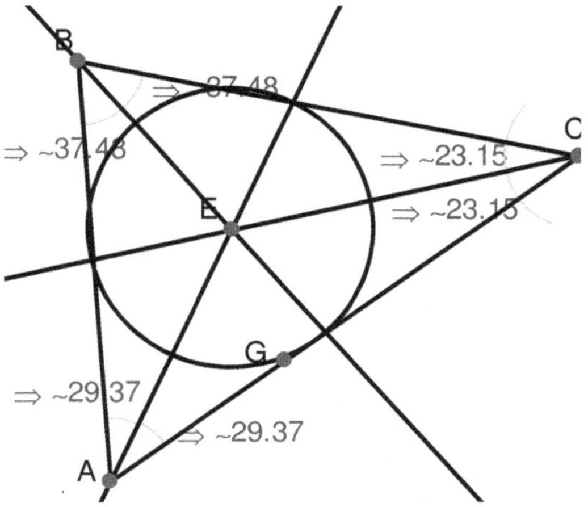

The diagram shows the intersecting angle bisectors meeting. The Incircle is also shown.

To draw the diagram

Draw the triangle

Draw the three angle bisectors
    Construct | Angle Bisector

Mark the intersection point
    Select any 2 of the bisectors

    Construct | Intersection 

Draw the circle with center at the intersection point.

Alternatively check the distance of the incenter from each side of the triangle.
    Select the Incenter and a triangle side
    Calculate | Distance/Length

Circumcenter

To create the drawing

Draw the triangle

Draw perpendicular bisectors
- Select a side of the triangle
- Construct | Midpoint
- Select the side and the midpoint
- Construct | Perpendicular

Mark the intersection

Draw the circumcircle
- Draw | Circle
- Click on intersection point and drag to one vertex of triangle.

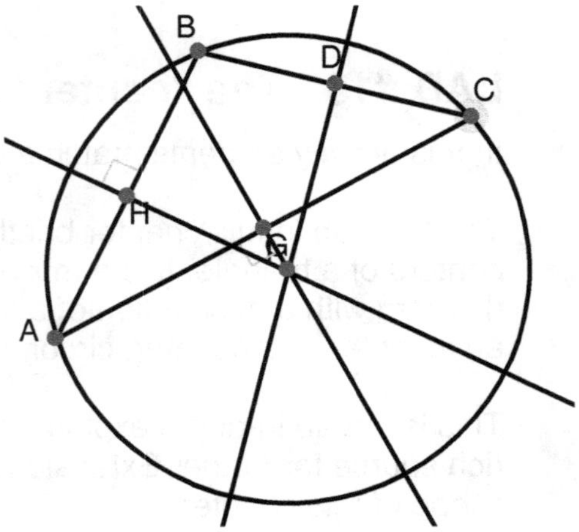

Centroid

Draw the triangle

Draw medians
- Select side
- Construct | Midpoint
- Draw | Line Segment between the midpoint and it's opposite vertex

Orthocenter

Draw the triangle

Draw altitudes
- Select a side and it's opposite vertex
- Construct | Perpendicular

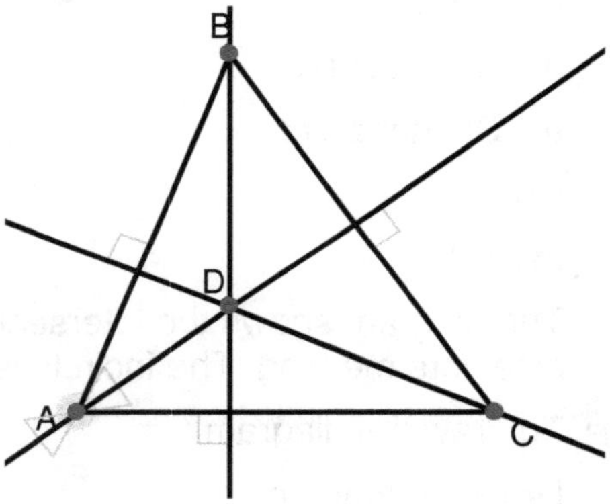

Steiner Point

Draw triangle

Place a point in the triangle

Measure the distance to each vertex
- Calculate | Real | Distance

Sum the distances
- Draw | Expression

Drag the point to minimize the distance

Measure the angle

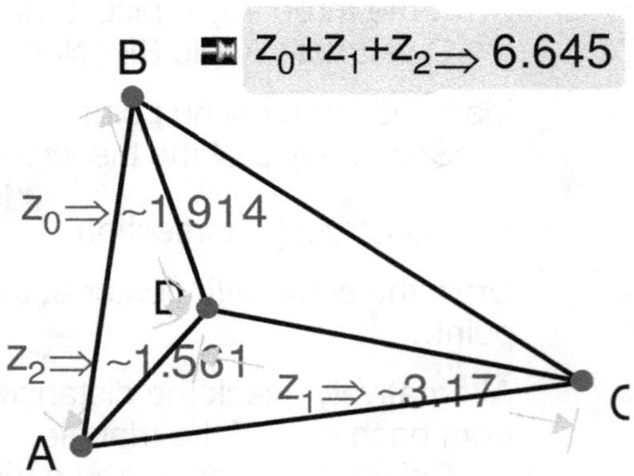

Chapter 7 – Slide, Turn, Flip, and Resize

Teacher Notes

The lab activities in this chapter cover the common transformations. A useful introduction can involve use of drawing software such as comes with most word processing programs. Such an approach ties in with students' intuitive understanding of transformations. For each transformation the first lab involves experimenting with the transformation and describing it in sentence form. These aim to develop a strong sense of the transformation, the ability to describe it and the parameters required to describe the transformation.

A follow up activity develops rules relating to specific transformations such as a quarter turn. The approach used is for students to gather numerical data, observe patterns developing and then generalize. The software has several features that support the investigations. The mathematical parameters required to define the transformation are explicitly required to create the transformation. For example a rotation requires a center and an angle. Also the ability to constrain aspects of the drawing in particular the coordinates of points makes it easy to experiment in a controlled fashion. For the more advanced student, the use of symbolic output makes the formulas explicit and provides a useful check on the preparatory work.

There is much that can be extended from these activities and it is hoped that users will consider such questions. Some starters include

- rotations by any angle;
- translations of the center away form the origin;
- glide reflections;
- translation of elementary functions and
- methods to determine the parameters of the transformation.

Labs in this chapter involve

- Describing translations in terms of vectors.
- Climbing up – An animation involving translations.
- Sliding with coordinates – A guided discovery of translation equations.
- Turn right – Describe rotations and identify.
- Turning with coordinates A guided discovery of equations relating to quarter turns about the origin.
- Ferris Wheel – An animation involving rotation.
- Flipping over – Describe reflections.
- Multiple flips – Explore multiple reflections.
- Kaleidoscope – Exploration of symmetry involving multiple reflections.
- Flipping and coordinates A guided discovery of equations for reflections in the axes and lines $y = \pm x$.
- Resize – Describe dilations.
- Resize and coordinates – A guided discovery of equations for dilations centered on the origin.
- Determining the Dilation – Determine the center and scale factor from an original figure and its dilated image.

LAB # 14 *Slide*

Aim: Explore and describe translations.

Draw a figure, perhaps something like a person.

Save your file so you can use it for other labs in this chapter.

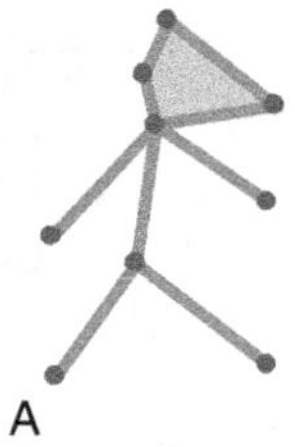

This figure was drawn using the line segment tool and then selecting the four segments of the head and creating a polygon. The color and line thickness can be changed before you start – Edit | Settings | Geometry.

Draw | Vector to represent the distance and direction of the translation. (note: a vector is a specified distance in a specified direction and is represented in GX as an arrow)

Select the whole figure

Use the Select tool and drag a rectangle to include all of the figure.

Construct translation

Click on Construct | Translation

Click on the vector

or, if you didn't already make the vector

Click and drag to create a translation vector. Release the mouse button to see the figure copied.

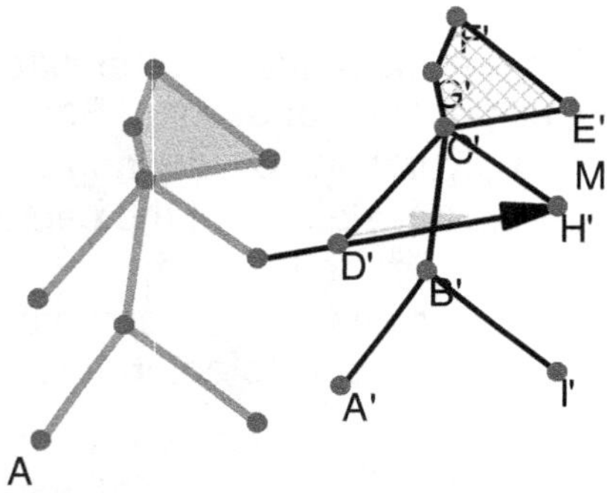

Drag the tip of the arrowhead on the translation vector. What happens to the image?

____

Drag the other end (start) of the translation vector. What happens to the image?

____

You have been exploring translations. Describe a translation in your own words. Think about what stays the same and what changes.

____

____

____

LAB #14 Slide

In this activity students

- Draw a stick figure using line segments and polygons;
- Describe translation in their own words and
- Use vectors to specify a translation

This activity is a short introduction to describing translations. The use of a stick figure may be appealing. The head of the figure in the diagram was created using the Construct | Polygon tool . It is likely that students will need to be shown this tool if they are to use it. The Draw | Polygon tool also works well for the head.

Middle school students are likely to be able to successfully complete this activity as well.

The questions are leading towards a description of translation as a slide, that is preserving size and orientation but allowing position to change and to describe this in terms of a vector.

Solution

The position of the image changes.

The position of the image changes. As the point is dragged to the left the image moves to the right.

The translated figure's size, shape and orientation remain the same (except the image of the unconstrained line that the vector is attached to in this example – but the vector doesn't have to be attached to the original). The position changes.

LAB # 15 *Climbing up!*

Aim: To further understanding of translations through animation.

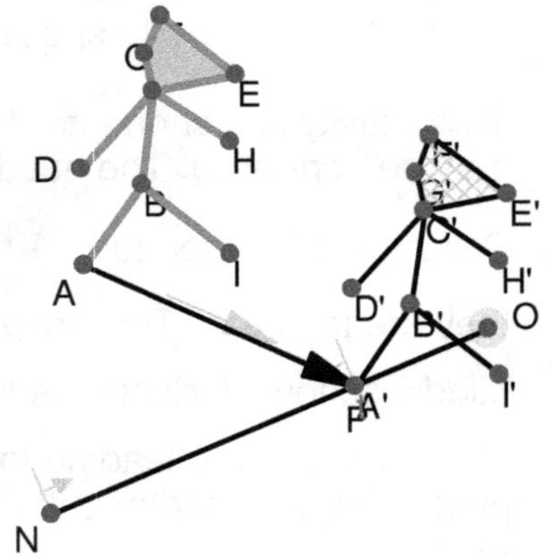

Draw a figure or open your saved file.

Draw a line segment to represent the hill.

Draw a point on the line segment.
- Draw a point.
- Select the line segment and the point.

- Constrain | Point proportional 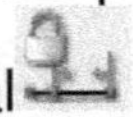.
- Accept the parameter value t by pressing enter.

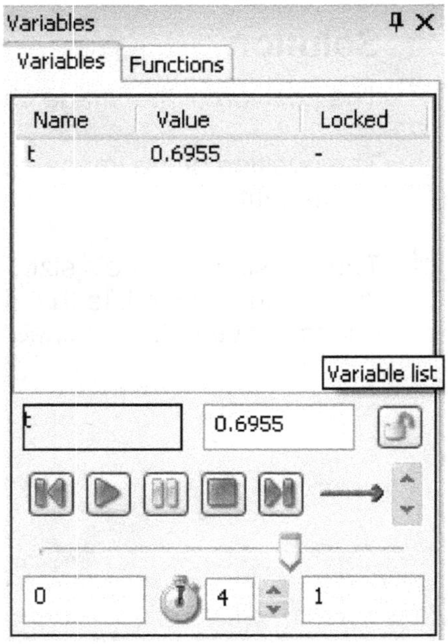

Move the point
- Go to the variables window of the Tools menu.
- Select the variable *t*
- Drag the slider and observe the point moving along the hill.

Animate the point
- Select *t* in the variables window
- Press play 
- The point should move to the end of the line.

Experiment with the different play modes by Clicking the up and down arrows.

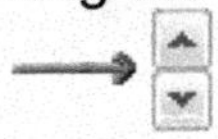

Construct a translation of the figure
- Select all of your stick figure - click and drag the selection box over the figure.
- Draw the translation vector from one foot to the point on the line segment.

Animate the figure.
- Select *t* in the variables window
- Press play 
- Enjoy!

EXTENSION

List the different animation modes and give a brief description.

Make the figure into a skier going down the slope.

LAB #15 Climbing up

In this activity students:

- Animate the stick figure drawn in the previous lab and
- Explore the animation controls in GX's Variables window.

This is a fun and engaging activity, resulting in a person or stick figure gliding up and down an incline. How to create an animation is the focus for this Lab. It is likely to take some time to understand the controls and capabilities contained in the variables window, however the effort is rewarding and once started students are likely to become involved in exploration. You may like to demonstrate aspects of the Variables window such as selecting the variable and pressing Play should your students need that support.

The activity builds on Lab #14 by using the same stick figure. (Instructions for creating the stick figure and constructing a translation of it are contained in the Lab *Slide*.) An incline is drawn, a point placed on it and constrained using the Point | Proportional tool, the point animated and the figure translated to this point.

There are many ways in which the Lab can be extended. The figure is restricted to a line, experiment with other shapes such as circles or equations. Constrain the translation as a vector $\begin{pmatrix} u \\ v \end{pmatrix}$ and then animate using u or v. Use vectors such as $\begin{pmatrix} u \\ u \end{pmatrix}$ and $\begin{pmatrix} u \\ 2u \end{pmatrix}$ etc.

To assign coefficients to a vector –

select the vector

Constrain | Coefficients 

Enter the values in the data entry box.

Animation is a valuable context for considering transformations and advanced students could investigate the ways in which computer animation is controlled.

Solution

Extension:

Go to beginning, Play, Pause, Stop Go to End

The animation modes are changed with the up and down arrows:

 Runs the animation one time through the specified range

 Runs the animation continuously from beginning to end

 Runs the animation forward and then back to the beginning

 Runs the animation in a continuous loop

LAB # 16 *Sliding with coordinates*

Aim: Describe translations in terms of vectors.
Describe how to determine coordinates of image points.

Draw a figure or open your previously saved file from Lab 14.

Constrain the coordinates of a point on the original figure to a value such as (2, 3)
Constrain the translation vector to (6, -2)

Calculate | Real | Coordinates of the image point.

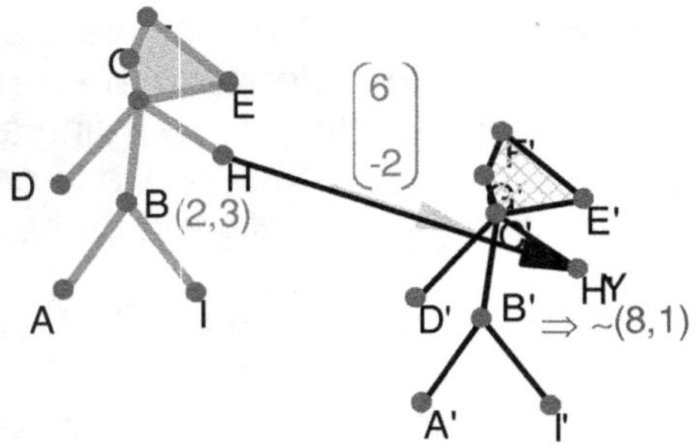

The diagram shows the point B with coordinates (2, 3) being translated by the vector to the image point B' with coordinates (8, 1).

If the coordinates of B were (5, 0) predict the image coordinates. (____ , ____)
Change the coordinates of B and check your prediction.

Repeat for other values.

Describe how to determine the coordinates of the image point from the coordinates of the original point and the translation vector.

____

Constrain B to (a, b) and the vector to (u, v)

The coordinates of the image point are (______ , ______).

Is this consistent with your findings? Explain. __________________________

____

EXTENSION

Write an algebraic statement to find the coordinates of object point where the translation vector is (u, v) and the image point is (c, d).

____

____

LAB #16 Sliding with coordinates

In this activity students:

- Place the stick figure on a coordinate system;
- Observe and record coordinates of object point and image point and
- Describe translations in terms of vectors

This is a guided discovery approach to the formulas for translating a point in terms of the horizontal and vertical components of the translation. The discovery begins with numerical values and by exploring new examples leading to the generalization encapsulated in the formulas.

The symbolic capability of GX enables the formulas to be displayed on the screen. You may like to show students this and discuss how the software might be generating this output.

Solution

If B is (5, 0) the image is (11 , –2)

The coordinates of the image point can be calculated by adding the first component of the translation vector to the x-coordinate and the second component to the y-coordinate.

If B is (a, b) and the translation vector is (u, v)
the coordinates of the image are $(a + u, b + v)$.

This is consistent with the statement above.

EXTENSION
$(c - u, d - v)$

LAB # 17 *Turn right*

Aim: To explore and describe rotations.

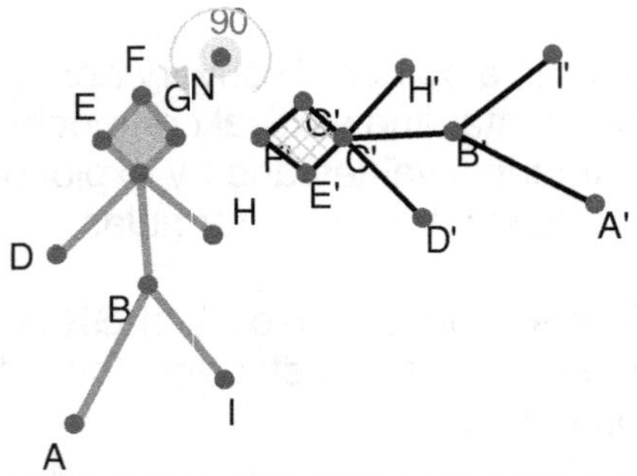

Create a figure.

Check that the angles are being measured in degrees.

Edit | Preferences | Math and set angle mode to degrees.

Rotate the figure 90°.

Select the figure.

Choose the rotation tool  .

Click a point on the screen.

Enter 90 in the input box.

Copy the drawing in the box.

Clearly show the center of rotation.

Draw a line segment between a point on the object and the center of rotation.

Draw a line segment between the corresponding image point and the center of rotation.

What is the angle between the line segments? ______°

Drag the center of rotation around.

Does this affect the angle between the line segments? ______________________

Repeat for a second object point.

Draw a line segment between the point on the object and the center of rotation.

Draw a line segment between the corresponding image point and the center of rotation.

Measure the angle between the two lines.

Investigate different angles of rotation.

Double click on the number and try the angles 180°, 45°, -90°, 135° and 270°.

Describe rotations in your own words. Think about what changes and what stays the same.

____

____

What information is needed to specify a particular rotation? ______________________

LAB #17 Turn right

In this activity students:

- Use Construct | Rotation tool;
- Rotate a stick figure and
- Articulate the factors needed to specify a rotation, center of rotation and angle of rotation.

The stick figure is used to emphasize the rotation of an object. The Rotation tool in GX is easy to use once the user understands the requirements; that is click on the screen to specify the center and then enter the angle. Students doing this lab are likely to work in degrees. Ensure the angle mode is set to degrees, Edit | Preferences | Math | Angle Mode.

The end point of the Lab is asking students what information is required to specify a rotation. This has been implicit throughout the Lab as students have been rotating their figure. They may need extra guidance to articulate that the center of rotation needs to be specified as well as the angle of rotation.

Solution

90°

Dragging the center of rotation doesn't change the angle between the line segments.

Rotations involve turning. The size and shape remain the same and the position and orientation change.

A point to rotate about and the angle turned through are needed to specify a particular rotation.

LAB # 18 *Turning with coordinates*

Aim: Explore patterns in coordinates with rotations of 90°, 180° and 270° about the origin.

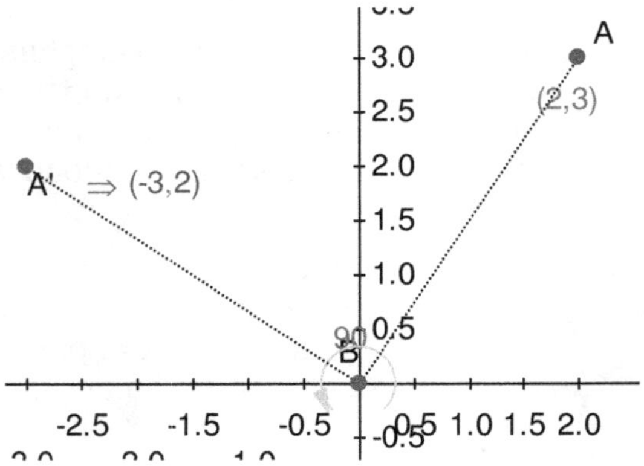

Open a new drawing in GX.

Display the axes

Draw the point (2, 3)
- Draw a point
- Select the point
- Constrain | Coordinates and enter 2,3

Rotate the point 90°
- Select the point
- Click the rotation tool
- Click on the origin
- Enter 90

Determine the coordinates of the image point
- Select the image point
- Calculate | Coordinates

Change the coordinates of the original point several times and complete the second column of the table.

Coordinates of point	Image coordinates clockwise 90° rotation	Image coordinates 180° rotation	Image coordinates anticlockwise 90° rotation
(2, 3)	(____ , ____)	(____ , ____)	(____ , ____)
(____ , ____)	(____ , ____)	(____ , ____)	(____ , ____)
(____ , ____)	(____ , ____)	(____ , ____)	(____ , ____)
(____ , ____)	(____ , ____)	(____ , ____)	(____ , ____)
(____ , ____)	(____ , ____)	(____ , ____)	(____ , ____)

Describe the pattern in the coordinates.

____

Change the rotation to 180 and complete the third column of the table.

Describe the pattern. ____

Change the rotation to 270 or -90 and complete the fourth column of the table.

Describe the pattern. ____

Summarize your findings by filling in the blanks:

(a, b) rotated 90° clockwise about the origin → (____ , ____)

(a, b) rotated 180° about the origin → (____ , ____)

(a, b) rotated 90° anti-clockwise about the origin → (____ , ____)

LAB #18 Turning with coordinates

In this activity students:

- Rotate the figure in the coordinate plane;
- Calculate and record coordinates of object point and image point and
- Develop formulas for the specified rotations about the origin.

This is a guided discovery approach to the formulas for the coordinates of an image after rotating about the origin by quarter and half turns. The discovery begins with numerical values and by exploring new examples leading to the generalization encapsulated in the formulas.

The symbolic capability of GX enables the formulas to be displayed on the screen. You may like to show students this and discuss how the software might be generating this output.

Solution

Coordinates of point	Image coordinates clockwise 90° rotation	Image coordinates 180° rotation	Image coordinates anticlockwise 90° rotation
(2, 3)	(–3 , 2)	(–2 , –3)	(3 , –2)
Answers will vary			
(x, y)	$(-y, x)$	$(-x, -y)$	$(y, -x)$
	The coordinates swap and the original y coordinate changes sign.	The coordinates swap and change sign.	The coordinates swap and the original x coordinate changes sign.

(a, b) rotated 90° clockwise about the origin → $(-b, a)$

(a, b) rotated 180° about the origin → $(-a, -b)$

(a, b) rotated 90° anti-clockwise about the origin → $(b, -a)$

LAB # 19 *Ferris Wheel*

Aim: Create a moving Ferris wheel using rotations and translations.

Open a new drawing in GX.

Draw a Ferris wheel
Draw a circle
You may like to add other features to make a more life like drawing.

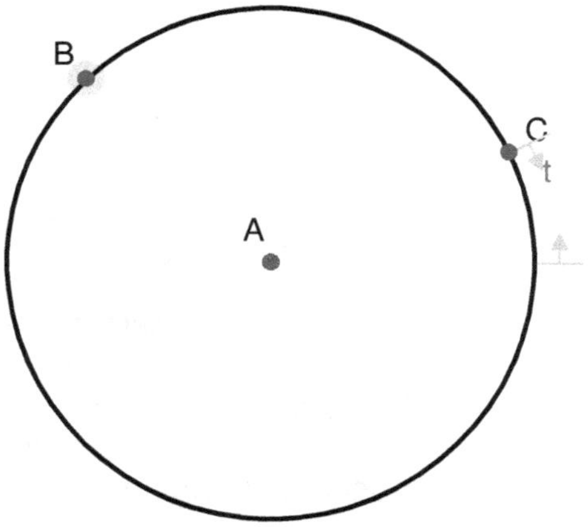

Point proportional on wheel
Place a point on the circle
Select point and circle
Constrain | Point proportional
Accept the default value for the parameter.

Animate point
Go to Variables window
Select the parameter
Press play

Construct points to hang gondolas from. Point C is one point.

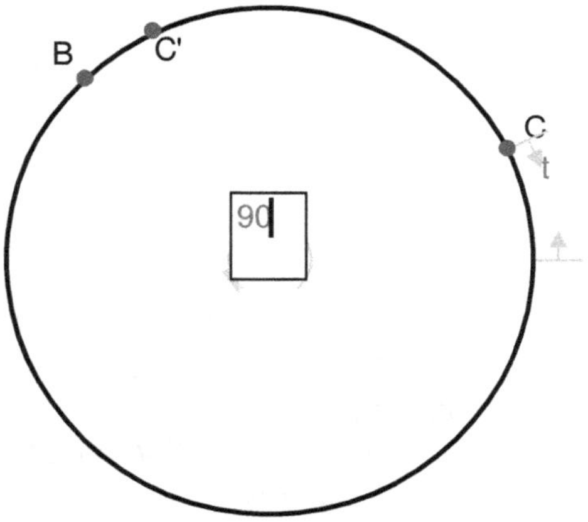

Place four of them around the circle.
Select the point (C)

Construct | Rotation
Click on the center of the circle (A)
Enter 90.
This creates point C'
(Note:
the angle measure should be in degrees
Edit | Settings | Math | Angle Mode)

Select the point C again
Construct | Rotation
Click on the center of the circle (A)
Enter 180.

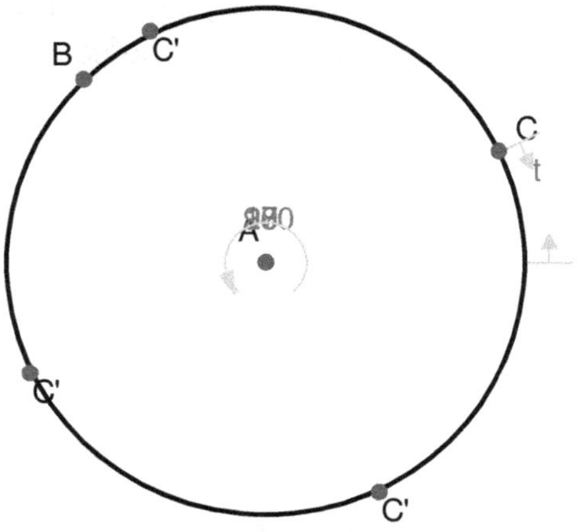

Select the point C again
Construct | Rotation
Click on the center of the circle (A)
Enter 270.

Draw a gondola

Draw connected line segments to make a gondola. Select these to construct a polygon (or just use Draw | Polygon ).

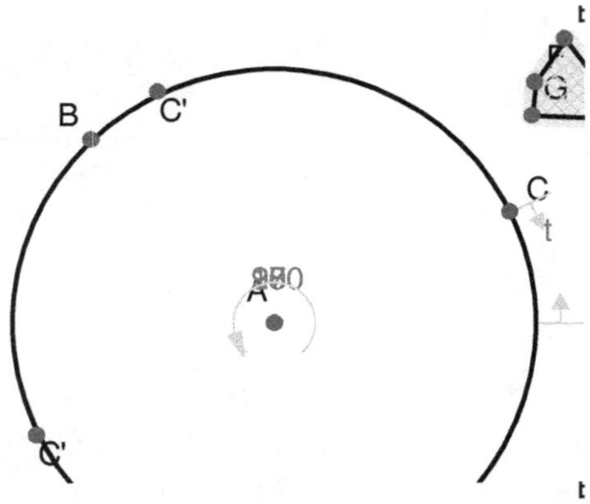

Translate gondola

Select the polygon

Construct | Translation 

Click on top of gondola and then on one of the four circle points.

Repeat for each point so you have four gondolas hanging from your Ferris wheel.

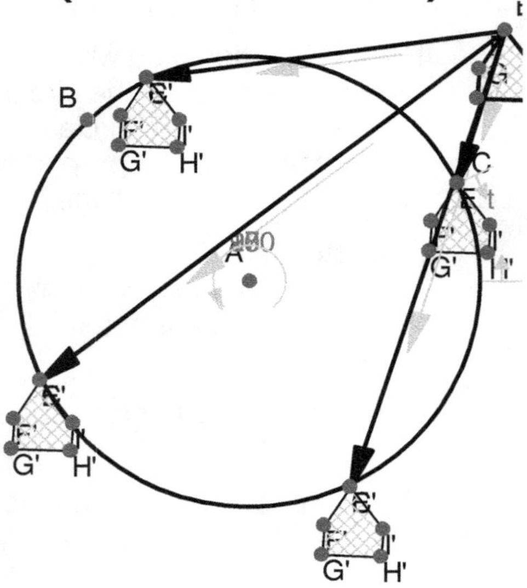

Hide the original gondola and labels you don't want to see.

Start the wheel

Set the animation mode to continuous

Press play

You can also change the properties of the polygons to get some color.

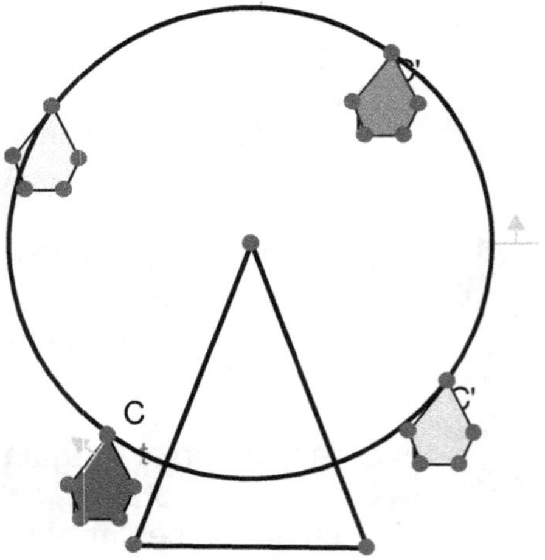

EXTENSION

Make a wheel with more equally spaced gondolas.

Create an animation for the Teacup ride.

LAB #19 Ferris Wheel

In this activity students

- Construct a diagram of a Ferris wheel in GX and
- Animate the Ferris wheel.

Animations are the focus of this activity. It combines the two transformations explored in this chapter so far to create a model of a Ferris wheel.

There is scope for extension such as creating models for other amusement park rides.
The Teacup ride is one such possibility. In the teacup ride the teacup spins at say 3 times the rate that the platform is spinning. The diagram shows a group of five teacups which is spinning as well, so the rider is experiencing the effect of three rotational motions.

Solution

Extension:
This is an outline of one method for constructing a model of the teacup ride with two platters.

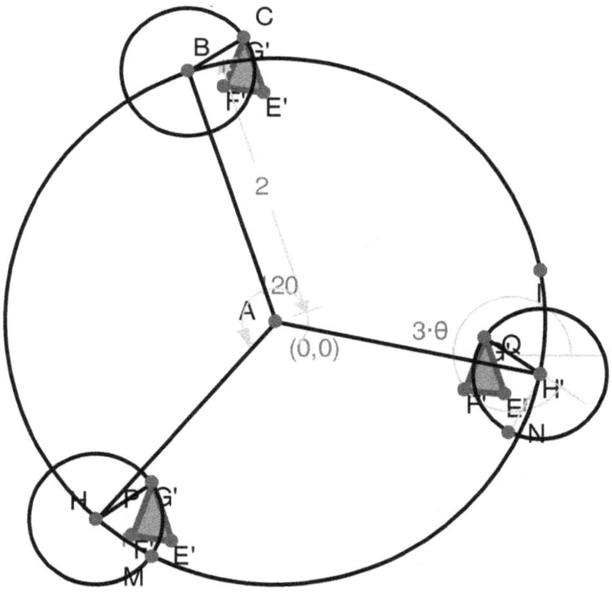

Construct | Circle
Constrain radius to r
Construct | Line Segment from the center of the circle to the circumference
Constrain the angle between the line and the x axis to θ
Construct | Rotation of Point, for three points rotate 120°
Construct circles centered on each point
Constrain radius to r / 3
Construct | Line Segment from the center of the smaller circles to the circumference
Constrain the angle between the line and the x axis to θ * 3
Rotate end of line segment for multiple teacups
Create the gondola

Translate a point on the gondola to each point on the teacup.
Hide the original gondola and translation vectors.

LAB # 20 *Flipping over*

Aim: Explore and describe reflections.

Construct a 3-4-5 triangle.

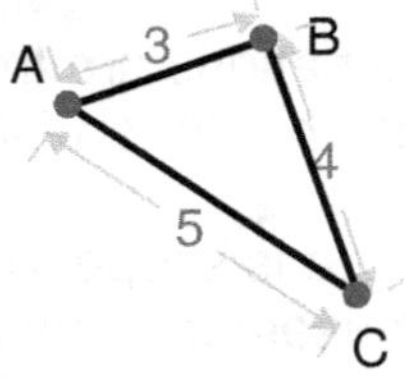

Draw | Polygon 

Click, move the mouse, click move again, click, move to the original point and click again.

Select tool, Click on one side (*e.g.* AB)

Constrain | Distance/Length 

Type 3 and press enter.

Constrain the other sides to lengths 4 and 5

Reflect in mirror line.

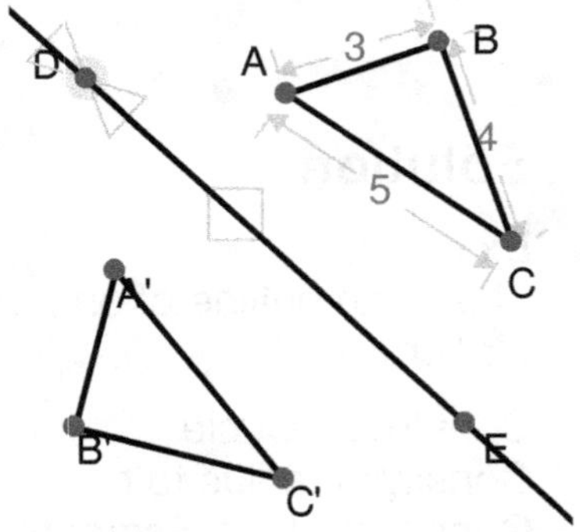

Draw two points.

Draw a line through the two points. This will be the mirror line.

Select the triangle.

Construct | Reflection 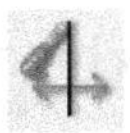

Click on the mirror line.

Save your drawing.

Drag the mirror line around so that the triangle crosses the mirror.

Sketch a copy of the screen in the adjacent box.

Draw line segments connecting vertices on the triangle to the corresponding image point, *i.e.* A to A' B to B' and C to C'

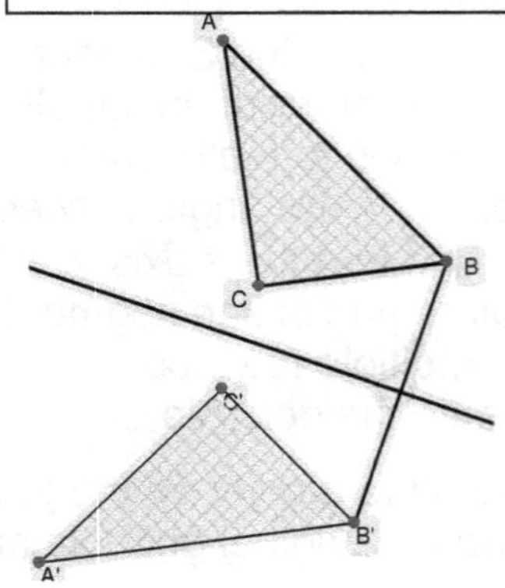

Delete the *Constraints* for the side lengths.

Drag the points A, B and C around.

Write a statement about the relationship between a line joining an object point to its image point and the mirror line?

Draw the reflection in the line drawn of the figure below.

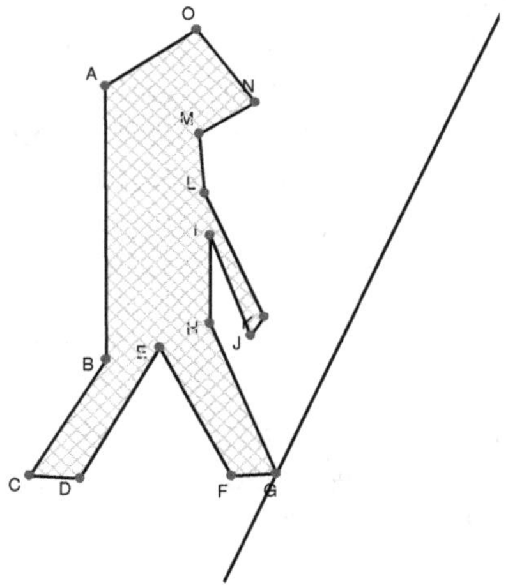

Create a similar drawing in GX. You can use the polygon tool .
Reflect it to check your work above.

Describe reflections in your own words. Think about what changes and what stays the same.

____

____

What information is needed to specify a particular reflection? ____________________

____

LAB #20 Flipping over

In this activity students:

- Are introduced to reflections;
- Use GX's Construct Reflection tool;
- Explore the property - the mirror line as perpendicular bisector of the line segment joining a point and its image point;
- Draw a reflection of a person on paper and compare this to a GX drawing and
- Describe reflections in their own words.

There are a number points about using the Construct | Reflection tool, that you may like to point out during the activity:

- You can create the mirror line first and select it during the Reflection operation, or you can create it during the selection step. The selected mirror line can be any line segment in the drawing (including a side of a polygon) or an infinite line. (Line segments are treated as infinite lines for the purposes of the reflection.)
- The selected mirror line is marked with a diamond (shown to the right).

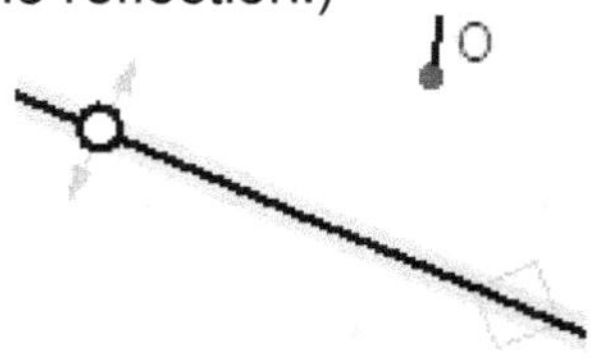

- Drawing | Infinite Line  - is a two step process:
  - Click once to place the line
  - Move the cursor to another point on the line and click-and-drag to rotate
- To reposition an infinite line:
  - Translate the line - click-and-drag
  - Rotate - select the line, click-and-drag another point on the line (away from the circle shown to the right)
  - Translate again (when the line is selected) – click-and-drag the circle on the line

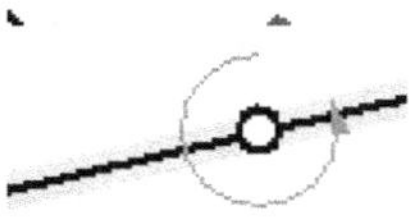

Solution

The line joining an object point to its image point is perpendicular to the mirror line. The mirror line bisects the object to image point line.

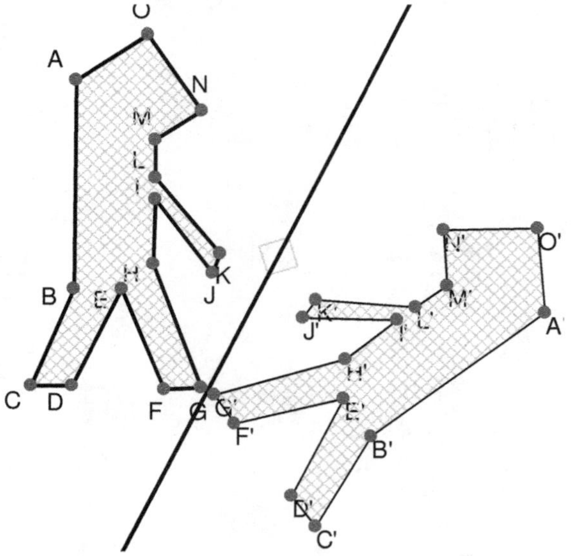

The size and shape of the figure remain the same. The position and handedness of the object change.

A line specifies the reflection.

LAB # 21 *Multiple flips*

Aim: To explore multiple reflections.

Open the drawing from the previous lab.

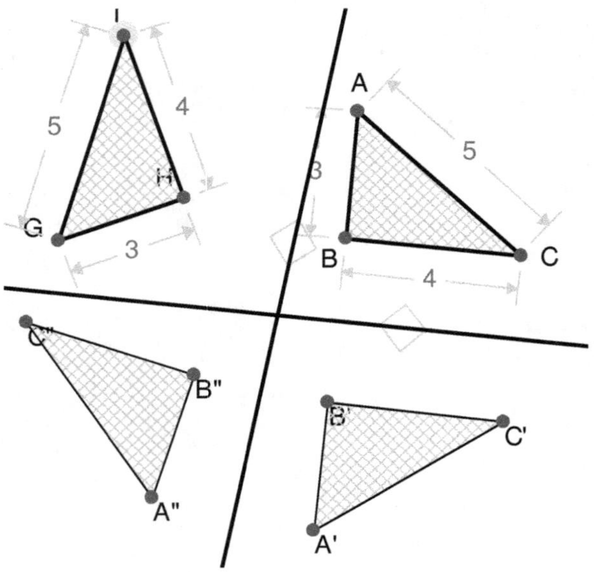

Create a second mirror line.

- Reflect A'B'C' in a second mirror line.
  Use the polygon tool to draw a triangle

Create a second 3-4-5 triangle (GHI in the figure to the right)

- Draw a triangle
- Constrain the sides to lengths 3, 4 and 5 units.
- Select the three sides and use the polygon tool to construct the triangle.
- By dragging and turning the new triangle, try to place the triangle on top of the original triangle and with each of the reflected triangles.

Note: Selecting the triangle and dragging will slide the triangle, selecting a vertex will rotate the triangle.

Which of the triangles ABC, A'B'C' and A"B"C" can you make GHI coincide with?

____

Explain your findings.

____

____

____

LAB #21 Multiple flips

In this activity students:

- Reflect triangles in two mirror lines and
- Explore which images match a copy of the object.

Two reflections are equivalent to a rotation; that is parity is restored. For extension students could locate the center of rotation and rotation angle equivalent to the double reflection. Beginning with mirror lines that are horizontal and vertical may be the easiest starting point.

The Lab is quite short, important to support Lab #22 and an opportunity to discuss "handedness". For example stereo isomers are molecules that are 3D mirror images. Some synthetically produced compounds have a different ratio of these isomers to naturally produced ones.

Solution

Triangles ABC, and A"B"C" can be slid and turned to coincide with triangle GHI.

Triangle A'B'C' is of opposite handedness or parity.

LAB # 22 *Kaleidoscope*

Aim: To explore multiple reflections as occurs in the Kaleidoscope.

Draw two line segments AB and AC

Construct triangle DEF using the polygon tool.

Construct triangle AED.

You might like to color the triangles too

- Select a triangle.
- Right click.
- Choose properties.
- Change the fill colour.
- Set the fill style to solid.

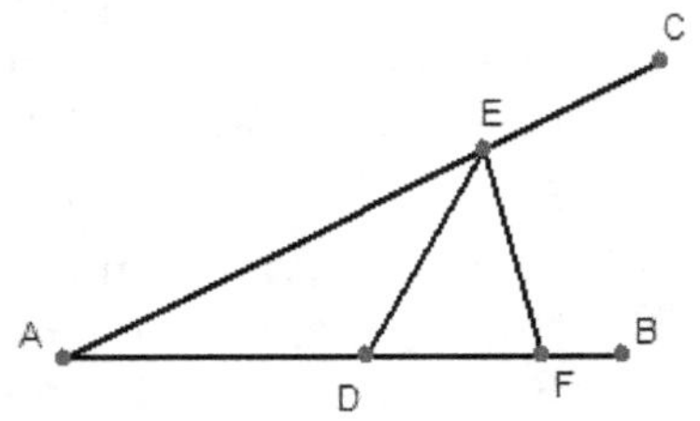

Select the triangles (Ctrl click for multiple selection) and construct reflection using AB as the mirror line.

Drag B and C around and play with the shape.

Constain E to the default values.

Can you drag E' to change the shape?

Explain. ________________________________

____

____

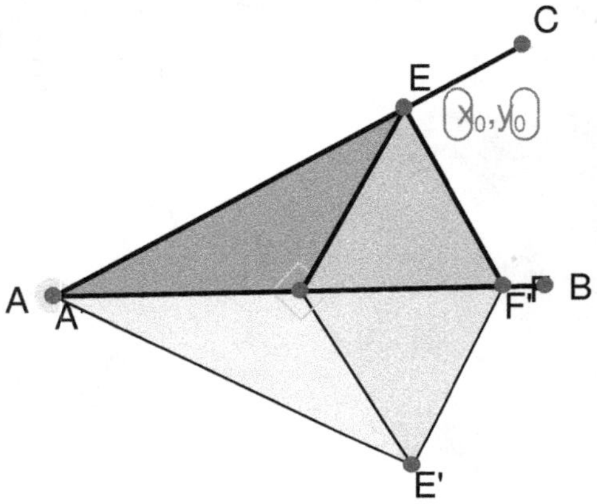

Reflect the 4 triangles in AE'

- Draw in the line segment AE'
- Select all four triangles
- Click the reflection tool
- Click on AF'

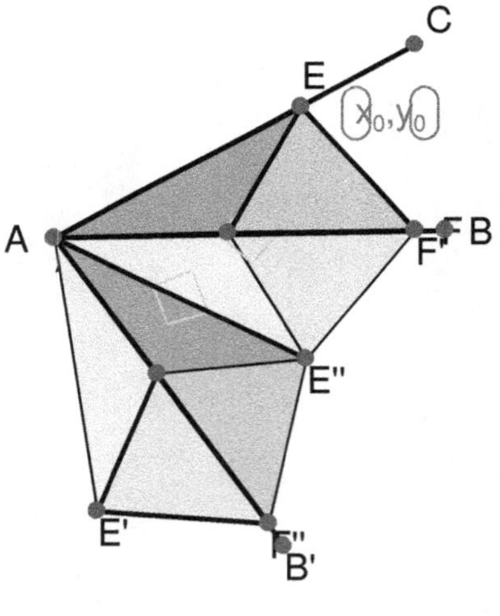

Add another 4 triangles to the figure with a further reflection. (this makes a total of 12 triangles)

Drag B or C to make the extreme edges line up.
In the drawing C is dragged until E' and E" are coincident.

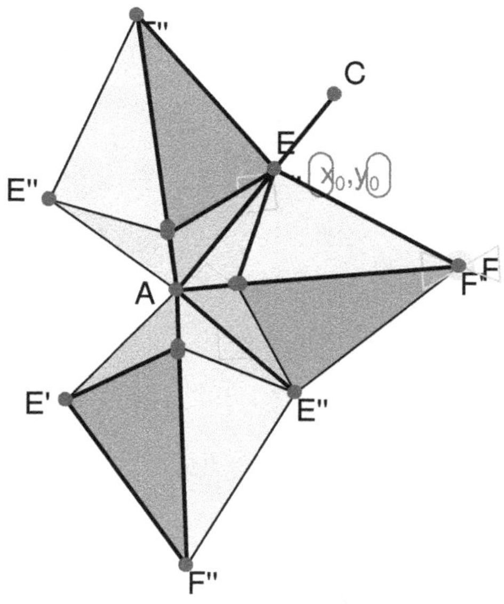

Describe the symmetry in your drawing. (Think about axes of symmetry and points of symmetry.

____

____

____

____

____

EXTENSION

The drawing shows one pattern created by dragging the points from the prvious sketch. Because there is overlap of the image points, you may have to Hide points to drag the desired points underneath.

Create your own interesting pattern. Copy the figure and save as an image file.

Add another 4 or 8 triangles by reflecting along one of the edges. i.e 16 or 20 triangles in total.

Drag B to make the extreme edges line up.

Constrain angle BAC to make the extreme edges line up.

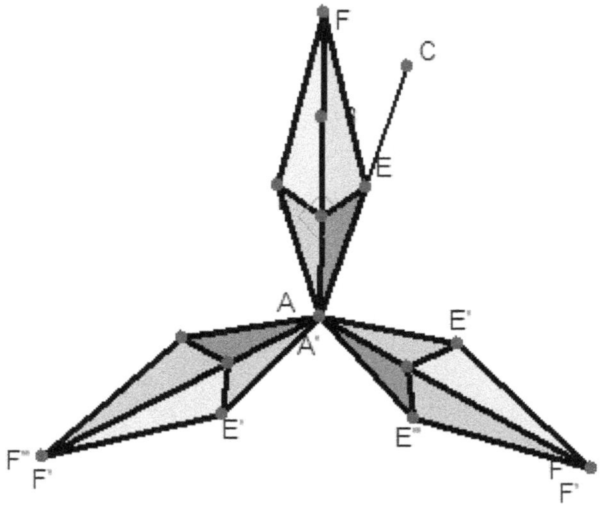

What symmetry does your figure now have?

Make a sketch …

LAB #22 Kaleidoscope

In this activity students:

- Create multiple reflections;
- Explore degrees of rotational symmetry and
- Have the opportunity for artistic expression producing a kaleidoscopic image.

The **kaleidoscope** is a tube of mirrors containing, loose colored beads or pebbles, or other small colored objects. The viewer looks in one end and light enters the other end, reflecting off the mirrors. Typically there are two rectangular lengthways mirrors. Any arbitrary pattern of objects shows up as a beautiful symmetric pattern because of the reflections in the mirrors. (Wikipedia 2007)

Toy kaleidoscopes are readily obtained from toy stores and would serve as a great addition or introduction to the activity.

It is intended to be a creative exercise. Student work can be captured through Edit | Copy drawing and then pasted into a word processor document. A description could be added and the resulting print outs (preferably in color) could then be displayed as a class project. An alternative would be to create a web page with the class images.

Changing the properties of the triangles is necessary to make an interesting or pleasing image. To do this right click on an object and select Properties. This screen appears.

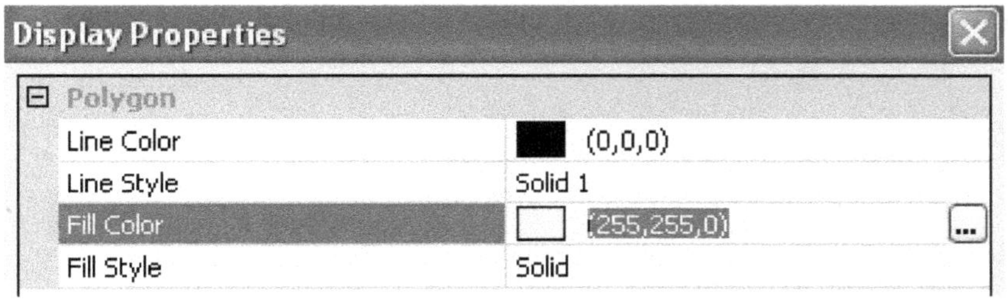

Select a property and then click the button on the right to change the color or style.

An interesting Extension activity would be to use animation to create a more realistic model of the kaleidoscope. Use the ideas of the Ferris wheel project to rotate some colored objects and then combine this with the multiple reflections.

Solution

You can not drag F' to change the shape. This because F' is determined by the position of F and the mirror line. Moving these changes the position of F'. Since there is a lot of overlap, you will need to use the Hide function to drag the desired points.

An image like that on the right has three lines of symmetry.

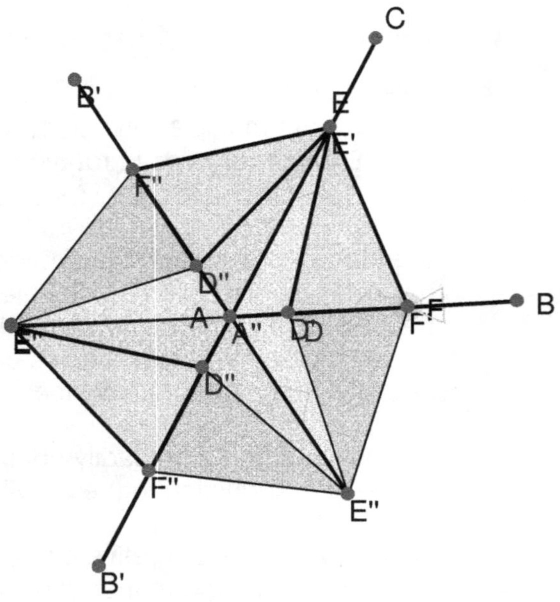

LAB # 23 *Flipping and coordinates*

Aim: To describe the effect of reflections in the axes and lines $y = x$ and $y = -x$ on the coordinates of a point.

Images in the *y* axis

Construct a triangle.

Construct a reflection of the triangle.

Constrain the first mirror line to be vertical (at an angle of 90° or $\pi/2$ if the angle settings are in radians)

Constrain one vertex of the triangle to a specific point *e.g.* (1, 4)

Show the axes.

Use the Calculate toolbox to find the coordinates of the reflected image of this point.

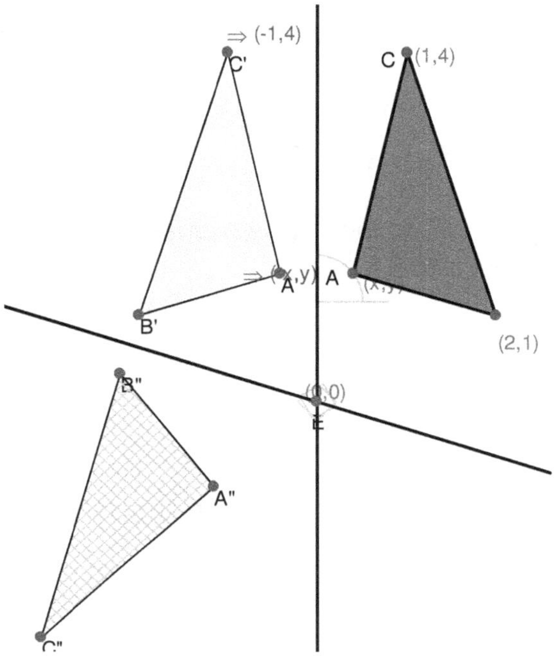

The image of (4, 1) reflected in the y-axis is (____ , ____)

Constrain a second vertex to another specific value *e.g.* (2, 5).

The image of (2, 5) reflected in the y-axis is (____ , ____)

Constrain the third vertex to a general point (x, y)

The image of (x, y) reflected in the y-axis is (____ , ____)

Reflection in the *x* axis:

Select the mirror lines angle constraint and change it to 0°

The image of (4, 1) reflected in the x-axis is (____ , ____)

The image of (2, 5) reflected in the x-axis is (____ , ____)

The image of (x, y) reflected in the x-axis is (____ , ____)

Describe in words what happens to the coordinates of a point when it is reflected in one of the axes.

Present an argument to justify that your findings must always be true.

____

____

Reflection in $y=x$

Select the mirror lines angle constraint and change it to 45°

The image of (4,1) reflected in the $y = x$ is (____ , ____)

The image of (2,5) reflected in the $y = x$ is (____ , ____)

The image of (x, y) reflected in the $y = x$ is (____ , ____)

Describe in words what happens to the coordinates of a point when it is reflected in the line $y = x$.

____

____

Reflection in $y = -x$

Select the mirror lines angle constraint and change it to –45° or 135°

The image of (4, 1) reflected in the $y = -x$ is (____ , ____)

The image of (2, 5) reflected in the $y = -x$ is (____ , ____)

The image of (x, y) reflected in the $y = -x$ is (____ , ____)

Describe in words what happens to the coordinates of a point when it is reflected in the line $y = -x$.

____

____

EXTENSION

Explore the coordinates of the image point for mirror lines at $x = h$

____

Explore the coordinates of the image point for mirror lines at $y = k$.

____

Explore mirror lines passing through the origin at other angles.

Multiple mirrors

Arrange two mirrors so that 4 reflections return to the original image.

How are the mirrors arranged? ______________________________

Arrange 3 mirrors so that 6 reflections return to the original image.

How are the mirrors arranged? ______________________________

LAB #23 Flipping with coordinates

In this activity students:

- Rotate the figure in the coordinate plane;
- Calculate and record coordinates of object point and image point and
- Develop formulas for the specified rotations about the origin.

This is a guided discovery approach to the formulas for the coordinates of an image point under these reflections. The discovery begins with numerical values and by exploring new examples leading to the generalization encapsulated in the formulas.

The symbolic capability of GX enables the formulas to be displayed on the screen. You may like to show students this and discuss how the software might be generating this output.

The Extension encourages students to explore horizontal and vertical mirror lines other than those passing through the origin and multiple reflections.

Solution

Reflection in the y-axis: (4, 1) → (–4, 1), (2, 5) → (–2, 5) $(x, y) \rightarrow (-x, y)$

Reflection in the x-axis: (4, 1) → (4 , –1) (2, 5) → (2, –5) $(x, y) \rightarrow (x, -y)$

When a point is reflected in the x-axis, the x coordinate remains the same and the y coordinate changes sign. In the y axis the y coordinate stays the same and the x coordinate changes sign.

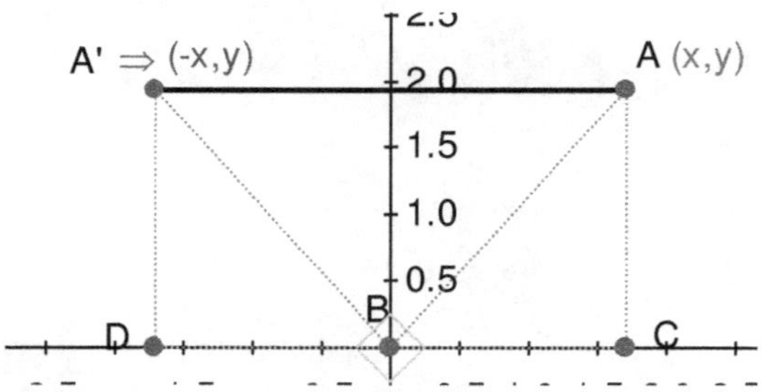

On the diagram the y coordinates are the same and the image point is the same distance to the left of the y axis as the original point is to the right.

Reflection in the line $y = x$: (4,1) → (1 ,4), (2,5) → (5, 2), $(x, y) \rightarrow (y, x)$
The coordinates swap.
Reflection in the line $y = x$: (4,1) → (–1, –4) (2, 5) → (–5, –2), $(x, y) \rightarrow (-x, -y)$
The coordinates swap and change sign.

Extension:
Reflection in the line $x = h$ $(x, y) \rightarrow (2h - x, y)$
Reflection in the line $y = k$. $(x, y) \rightarrow (x, 2h - y)$

Multiple mirrors

When two mirrors are at right angles 4 reflections return to the original image.
When three mirrors are at 60° and 120°, 6 reflections return to the original image.

LAB # 24 *Resize*

Aim: To explore and describe dilations.

Draw a figure or load your previously saved drawing of a figure from Lab #14.

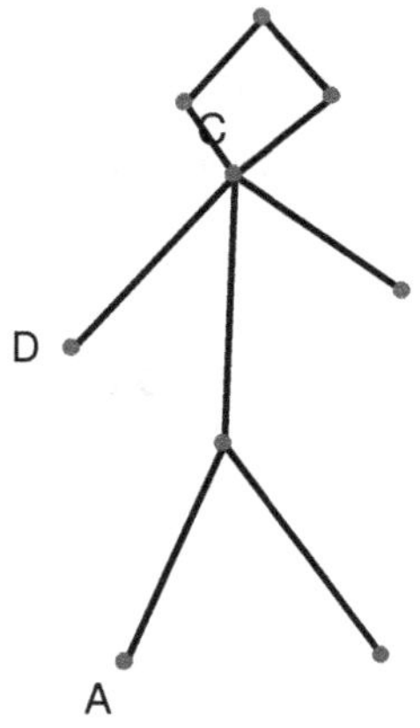

Select the whole figure
Select tool and drag a rectangle to include all of the figure

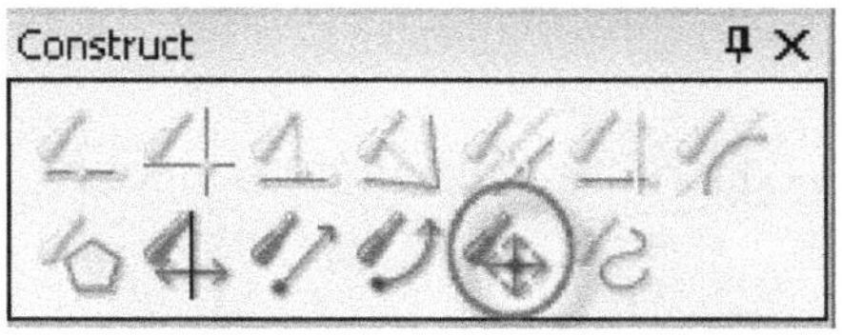

Construct | Dilation
Click a spot somewhere on the screen. (this is the center of the dilation)
Enter 2 (this is the scale factor)
You can drag the center of dilation around to make the effect clearer.

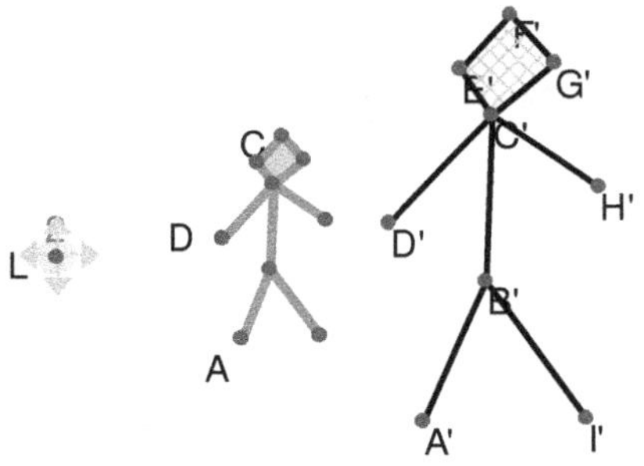

Describe the effect.

Change the scale factor to 3
Double click on the number next to the center of dilation

Describe the effect.

Experiment with other numbers, include decimals between 0 and 1 and negative numbers.

Make a statement about the effect of changing the scale factor.

Change the scale factor to *a*.

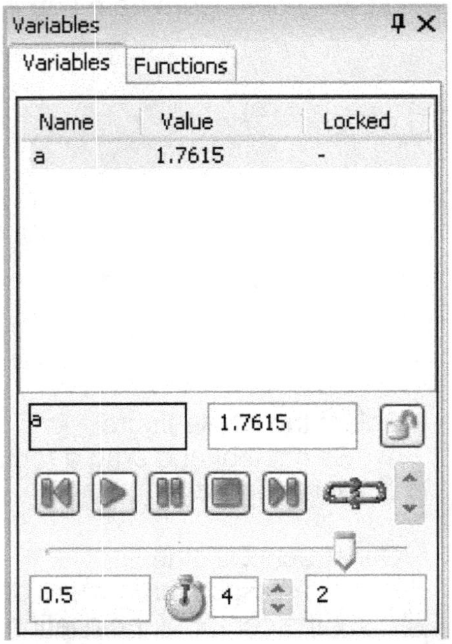

Go to the variables window.
Select the variable *a*

Use the slider to change the scale factor. 

Set minimum and maximum values for the scale factor.

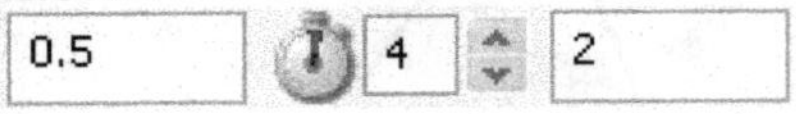

Run the animation. 

Run the animation as a loop 

Describe the effect.

Repeat after changing the range of values for *a* to –1.5 to 1.5

Describe the effect.

What is the difference between dilation and enlargement?

LAB #24 Resize

In this activity students:

- Are introduced to dilations;
- Use GX's Construct Dilation tool;
- Dilate a stick figure;
- Experiment with scale factors greater than1, between 0 and 1 and less than 0 and
- Animate the dilation.

There is a lot involved in completing this lab. Students who have worked through earlier labs in this chapter should be able to complete the activities in a single lesson. Some extensions using the animation can involve changing the boundaries for the enlargement variable (scale factor). For example Hide the original figure and simulate the figure disappearing into the distance. Simulate the figure approaching from a distance.

Some students may require prompting to explain that all enlargements are dilations but not all dilations are enlargements.

Solution

The figure retains the same shape and orientation. The size of the figure doubles.
Scale factor of 3: the figure is three times as large.
When the scale factor is between 0 and 1 the figure is smaller.
When the scale factor is negative the figure is inverted.
When the scale factor < -1 the figure is inverted and larger.
When the scale factor is between –1 and 0 the figure is inverted and smaller.
The figure gradually increases from half size to double size.

For *a* –1.5 to 1.5, The figure begins inverted. It then diminishes in size to zero and then grows in size, not inverted.

Dilation includes reduction and enlargement.

LAB # 25 *Resize and coordinates*

Aim: To develop formulas for the coordinates of a point when it is dilated about the origin.

Turn the axes on

Draw a point A

Constrain A to (0, 0)

Draw a second point B and constrain it to some small integral values such as (2, 3)

Dilate B using the origin (A) as the center of dilation

and scale factor 2.

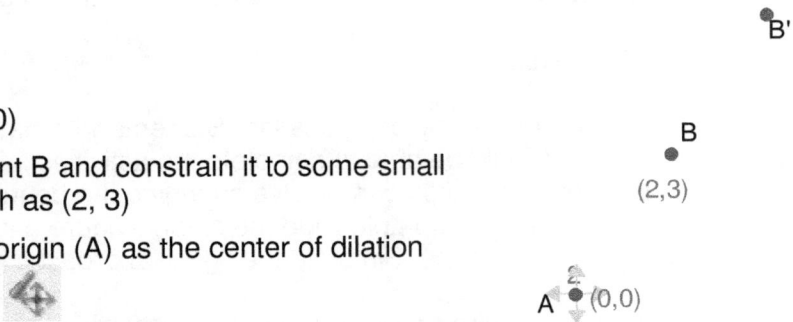

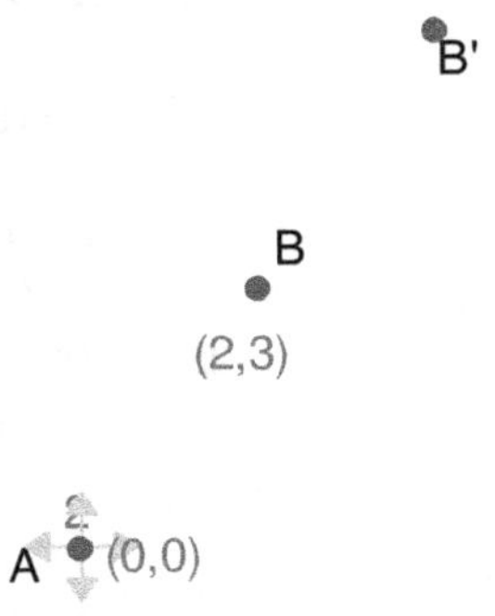

Display the coordinates of the image point B'. Calculate | Real | Coordinates

Change the coordinates of B and the scale factor to complete the table.

Coordinates of point	Image coordinates		
	scale factor = 2	scale factor = 0.5	scale factor = -3
(2, 3)	(____ , ____)	(____ , ____)	(____ , ____)
(____ , ____)	(____ , ____)	(____ , ____)	(____ , ____)
(____ , ____)	(____ , ____)	(____ , ____)	(____ , ____)
(____ , ____)	(____ , ____)	(____ , ____)	(____ , ____)
(____ , ____)	(____ , ____)	(____ , ____)	(____ , ____)

Describe the patterns you observe.

____

____

____

Summarize your findings by completing the statement

(a, b) —— dilation center (0, 0) scale factor k ——→ (____ , ____)

Note: You can check your answer by changing the *Constraints* to these variables.

LAB #25 Resize with coordinates

In this activity students:

- Dilate points with the origin as the center;
- Look for patterns in the coordinates of the image point and
- Generalize the patterns into a formula for the coordinates of the image point.

This is a guided discovery approach to the formula for the coordinates of the image point. The discovery begins with numerical values and by exploring new examples leading to the generalization encapsulated in the formula.

Solution

Coordinates of point	Image coordinates		
	scale factor = 2	scale factor = 0.5	scale factor = -3
(2, 3)	(4 , 6)	(1, 1.5)	(–6, –9)
(____ , ____)	(____ , ____)	(____ , ____)	(____ , ____)

Other answers will vary.

The coordinates are multiplied by the scale factor.

$$(a, b) \xrightarrow{\text{dilation center } (0,\ 0) \text{ scale factor } k} (ka, kb)$$

LAB # 26 *Determining the Dilation*

Aim: To determine the center and scale factor of a dilation from a figure and its image..

How can you determine the scale factor and center of a dilation from an image?

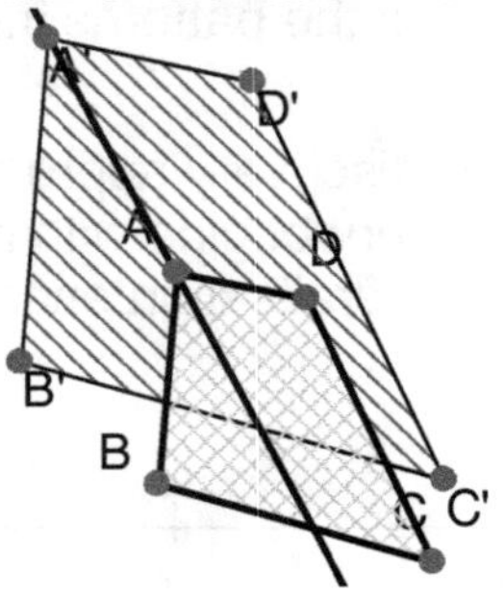

Create a quadrilateral.

Dilate it by the default factor a.

Hide the center.

Drag the figure(s) around.

Draw a line going through an object point and its image point.

Repeat for the other three points.

What do you notice? ______________________________

Show all to reveal the center of the dilation.

Calculate lengths from the center to the image and the centre to the object. But first- Edit | Settings | Math | Output and select True for Show Name

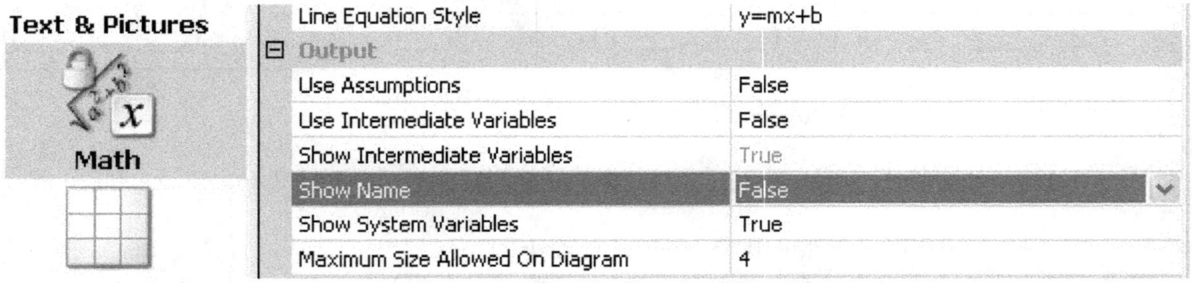

Create an expression to determine the ratio shown at right.
Calculate lengths

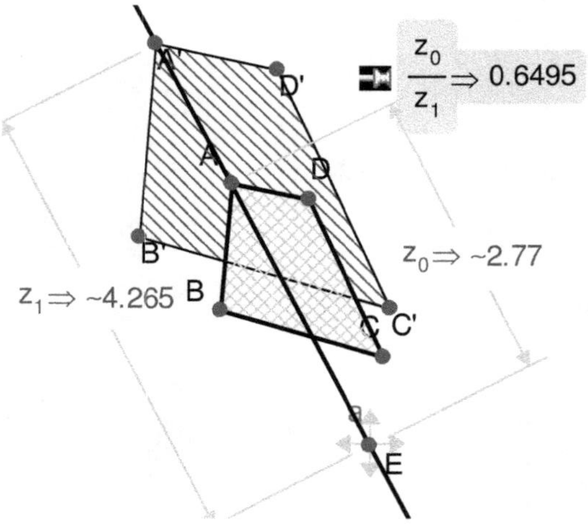

Draw | Expression
Use the Symbols menu to create a fraction

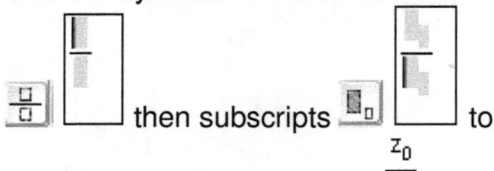

then subscripts to

complete the expression like $\frac{z_0}{z_1}$ and press enter.

Repeat for each object point and image point.

What do you notice about the ratio? ______________________________

____

Note: Look at the contents of the Values Window too.

LAB #26 Determining the Dilation

In this activity students

- Draw a quadrilateral and dilate it;
- Hide the center of the dilation;
- Find the intersection of lines passing through a point on the object and the corresponding image point and
- Calculate the ratio of the distance from the center to the image point and the center to the object point.

This is the first time in this book an expression has been used. In GX, the Expression tool is used to perform calculations that the user specifies. When GX calculates a quantity it uses a symbol to represent the quantity – "z" with a subscript number. If the symbol is hidden, change the default setting: Edit | Settings | Math | Output and select True for Show Names. It can also be done for an individual output by selecting the Output, right clicking and choosing Properties to change the Show Name toggle.

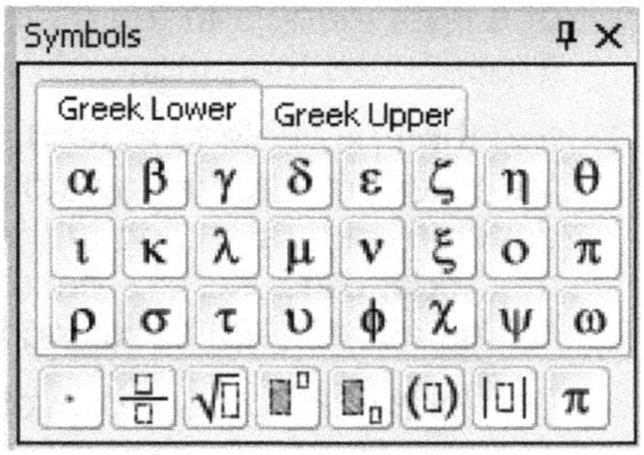

I found it took some time to get used to using subscripts, so it may be worth while to practice before trying this in class. The Symbols window is shown to the right. The bottom row has buttons to input fractions and subscripts.

The easy way is to use / for fractions and [ ] for subscripts, *e.g.* z[0] / z[1] will be displayed as:

$$\frac{z_0}{z_1}.$$

Direct students to look at the contents of the variable window to compare with their calculated values.

In summarizing the activity lines through an object point and corresponding image point meet at the center of the dilation and the ratio is the scale.

Solution

The lines pass through the same point, the center of the dilation

The ratios are equal and equal to the scale factor.

Chapter 8 – Right into Triangles

Teacher Notes

A common structure for teaching right triangle trigonometry is to ask students to memorize the rules for the trig ratios through a mnemonic such as SOHCAHTOA. For many students rushing to this doesn't give students the opportunity to understand either the historical development or an intuition for the ratios and their use.

The labs in this chapter attempt to emphasize the measurement aspect for developing right triangle trigonometry. The starting point is the measurement of triangles with a unit hypotenuse "standard triangles" and to then use similarity arguments to solve problems. Such an approach can enable students to appreciate what the trig ratios for angles between 0 and 90° represent.

GX makes it easy to measure right triangles. *Constraints* are the basis for this.

Extending beyond right triangles is achieved by showing the standard triangles in the first quadrant of the unit circle. This approach can support the transition from the ratios of a right triangle to the concept of circular functions. Clearly students need far more experiences than those offered in this chapter to make such a transition.

It is then possible to extend solving triangles to include non right angles triangles with the sine formula and laws of cosines. Playing with such triangles enables students to develop understanding of the sine and cosine rules as a precursor to pencil and paper methods. Students are encouraged to practice the pencil and paper methods and to use GX as a means of checking their work.

The labs in this chapter involve:

- Pythagorean theorem
- Converse of the Pythagorean theorem
- Right triangles - 30 and 45 degree triangles
- Measures of standard triangles
- Use similar triangles to solve contextual problems involving right triangles
- Convert Cartesian coordinates to polar coordinates
- Derive the Sine formula
- Derive the Law of cosines
- Solve four problems involving non-right angled triangles
- Solve a problem requiring use of sine Formula and law of cosines
- Use the given data or collect your own to make measurements at a distance.

The exercises conclude with target practice, investigations that involve changing coordinate systems and range finding.

LAB # 27 *Pythagorean theorem*

Aim: To explore the Pythagorean Theorem using GX.

Part 1 - Draw triangle

Set one angle to 90°
    Select the two sides
    Constrain angle to perpendicular

Set lengths of legs to 3 and 4.
    Select side
    Constrain | Distance / Length

What do you expect the length of the hypotenuse to be?

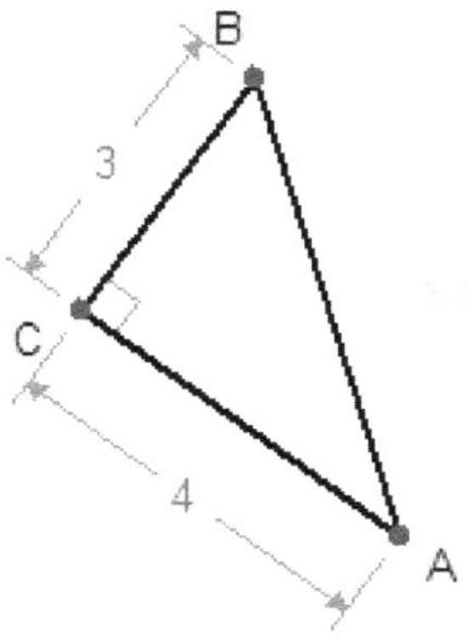

____

Measure hypotenuse
    Calculate | Real | Distance / Length
    (Toolbar note: make sure to use the Real tab.)

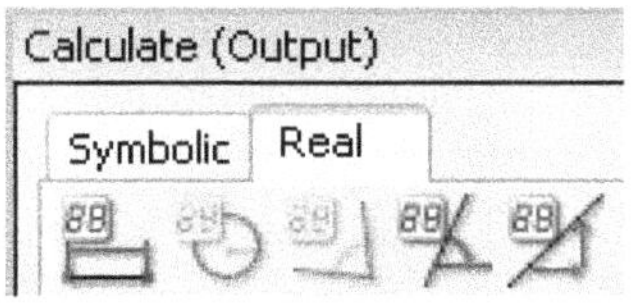

Experiment with other lengths of the legs
    Double click on the length and change the value

Record the leg lengths and the measure of the hypotenuse.

Leg 1	Leg 2	Hypotenuse
3	4	

Part 2 The formula

Constrain the leg lengths to a and b
    Double click on the length
    Change the value

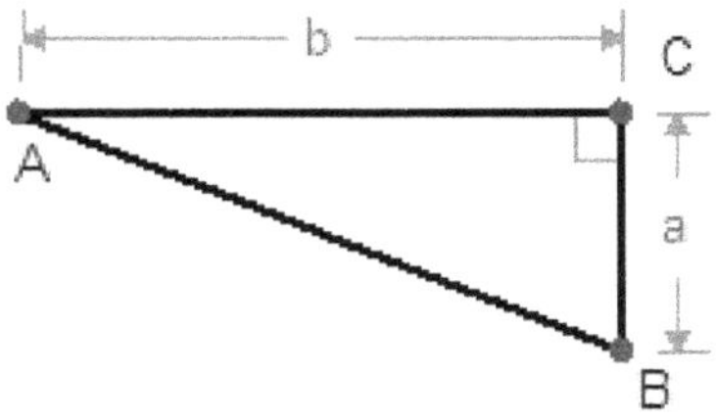

Measure the hypotenuse
    Calculate | Distance
    (Toolbar note: make sure that the Symbolic tab is used.)

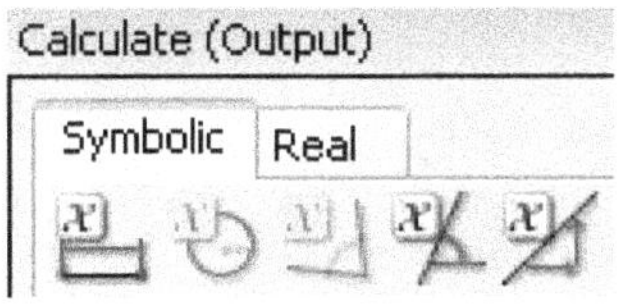

Length of hypotenuse AB = ______________________________

Part 3 Missing Leg

Set lengths of one leg to 5 and hypotenuse to 13.
Double click on the length and change the value
Predict the length of the other leg? ______

Measure hypotenuse
Calculate | Real | Distance/Length

Experiment with other lengths of the hypotenuse and leg.

Record your results.

Hypotenuse	Leg 1	Leg 2
13	5	

Part 4 Missing leg in symbols

Constrain the length of the hypotenuse to c

Constrain length of one leg to a

Measure other leg symbolically.

Length of third leg = ______________

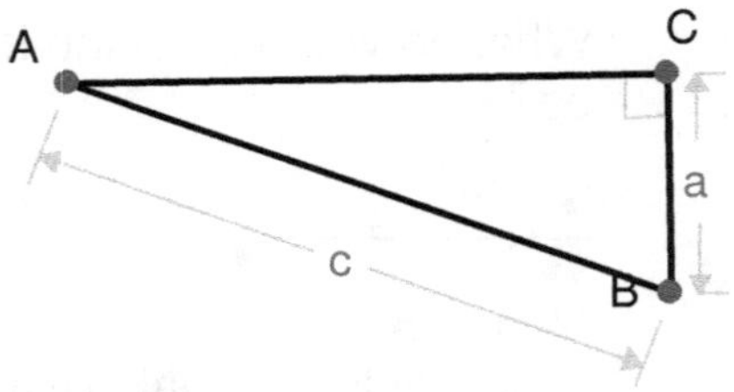

Is this the same as the formula you have used?

Explain

____

____

EXTENSION

Use GX to search for some Pythagorean triples.
Pythagorean triples are when all three lengths of a right triangle are integers.
You could use a guess and check approach.

LAB #27 Pythagorean theorem

In this activity students:

- Construct a right triangle;
- Use GX to measure the hypotenuse for a numerical examples;
- Measure symbolically to generate the formula;
- Measure right triangles with a missing leg numerically and
- Measure right triangles with a missing leg symbolically

This Lab is a suitable for students' first experience with GX. It is assumed that students have some familiarity with the Pythagorean Theorem and so the can focus on the software. To get angle measures in degrees ensure that GX's preferences are set to degrees. Edit | Preferences | Math | Math | Angle mode.

The use of *Symbolics* is designed to reinforce the algebraic representations of the Pythagorean formula commonly used. For students familiar and comfortable working with radicals, the use of real measurements is redundant. Directing your students to use symbolic measure enables them to get results expressed as radicals. These radicals in simplified form can lead to useful reinforcement of work associated with radicals.

Searching for Pythagorean triples is suggested as an Extension activity. A group or whole class utilizing a trial and error approach is envisaged.

Solution

Part 1

Leg 1	Leg 2	Hypotenuse
3	4	5

Part 2

AB = $\sqrt{a^2 + b^2}$

Part 3

Hypotenuse	Leg 1	Leg 2
13	5	12

Part 4

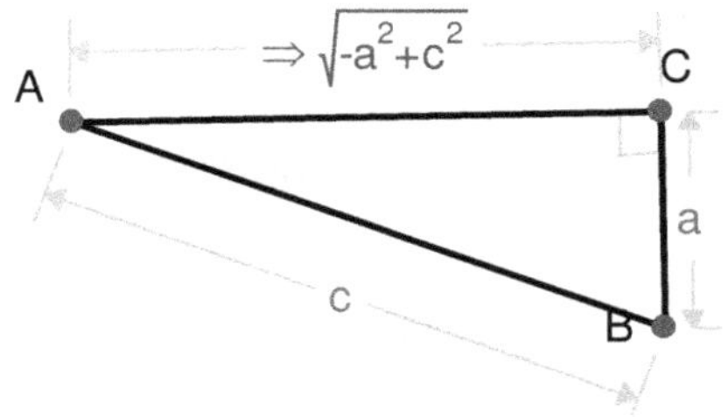

EXTENSION

Some examples of Pythagorean triples are {5, 12, 13}, {7, 24, 25}, {6, 8, 10}
In general for integers m and n; $\{|m^2 - n^2| : 2mn : m^2 + n^2\}$ is a Pythagorean triple.

LAB # 28 *Converse of the Pythagorean theorem*

Aim: To use the Pythagorean Theorem to determine whether triangles are right angled.

Part 1

Draw triangle ABC

Constrain the side lengths to 5, 12 and 13.

Measure the angles.

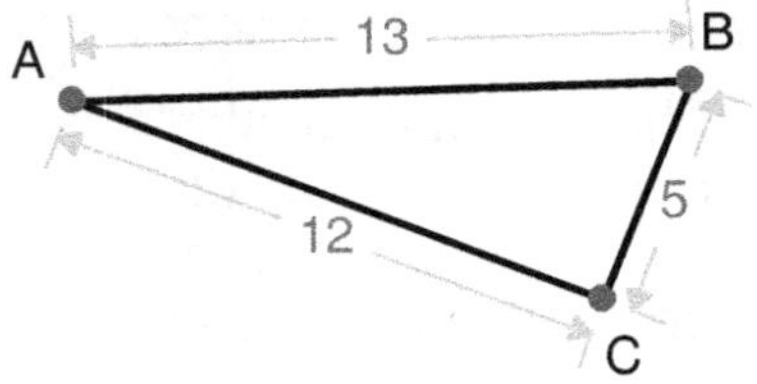

m∠A = ___________

m∠B = ___________

m∠C = ___________

Is this a right triangle? ______________

Use your drawing to decide which of the following triangles are right angled.

a	b	c	Right-angled?
5	6	7	
6	8	10	
7	24	25	
42	56	70	
11	30	31	
11	60	61	

$a^2 + b^2$	c^2

Calculate to complete the table.

Write a statement about how you can decide which triangles are right angled.

____

____

____

Part 2

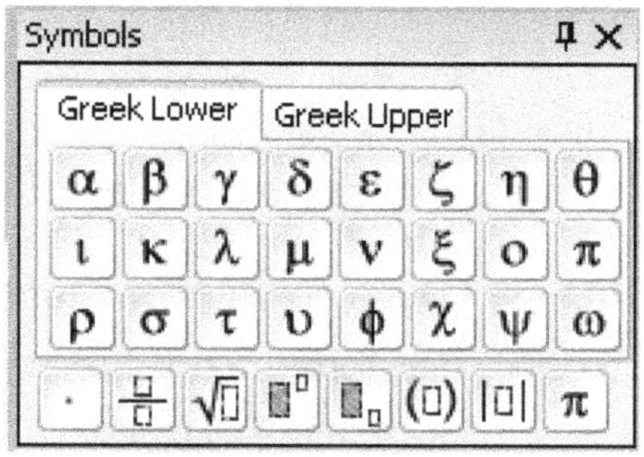

Draw triangle ABC

Constrain the side lengths to *a*, *b* and $\sqrt{a^2+b^2}$.

Use $\sqrt{\square}$ from the symbols palette of the toolbar to enter the radical expression.

Measure the angles.

m∠A = ____________

m∠B = ____________

m∠C = ____________

Is this a right triangle? ___________

Does it make a difference what the lengths of a and b are? _______________________

Drag vertices to test your predictions.

Repeat with lengths a, b and $\sqrt{a^2-b^2}$

Write a conclusion.

____

____

____

LAB #28 Converse of the Pythagorean theorem

In this activity students

- Construct triangles given the lengths of the three sides using *Constraints*;
- Determine which of these are right angled;
- Develop an understanding of the converse to the Pythagorean Theorem and
- Explore the positioning of the right angle in triangles with sides a, b and $\sqrt{a^2+b^2}$ and a, b and $\sqrt{a^2-b^2}$.

An exploration involving numerical examples in which the triangles look close to right angled triangles helps students to contextualize the theorem.
Part 2 involves *Symbolics* and can be extended with other expressions. The expressions for the non-right angles will be messy. You may suggest to students they omit those questions. The longest side of the triangle may not be obvious and the activity provides an opportunity to discuss the expressions and what can be derived, that is $a > b$ and $a > \sqrt{a^2-b^2}$.

Solution

Part 1

m∠A = 22.6°, m∠B = 67.38°, m∠C = 90° and the triangle is right angled.

a	b	c	Right-angled?
5	6	7	No
6	8	10	Yes
7	24	25	Yes
42	56	70	Yes
11	30	31	No
11	60	61	Yes

$a^2 + b^2$	c^2
61	49
100	100
625	625
4900	4900
1021	961
3721	3721

A triangle is right-angled if the sum of the squares of the legs equals the square of the longest side.

Part 2

m∠A = 90°, and m∠B and m∠C will vary. The triangle is right-angled.

No. The triangle is right angled for all values of a and b.

The triangle a, b and $\sqrt{a^2-b^2}$ is right angled.

LAB # 29 *Right triangles - 30° and 45°*

Aim: To determine and explore exact trig ratios.

Some right-angled triangles have sides in interesting ratios.

Construct a right triangle with hypotenuse 1 and angle 30°.
- Construct triangle
- Constrain C to 90°
- Constrain the length of the hypotenuse (AB) to 1
- Constrain angle A to 30 °

Measure sides BC and AC symbolically.
- Calculate | Symbolic | Distance / Length

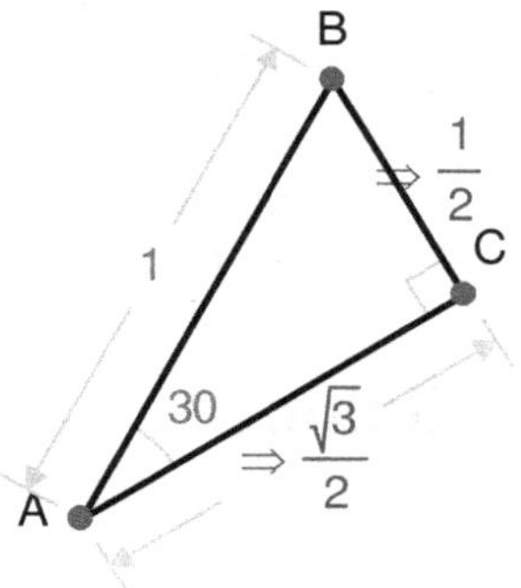

Change the value of A and complete the table.

Angle (A)	Leg 1 (BC)	Leg 2 (AC)
15		
30	$\frac{1}{2}$	
45		
60		
75		
50		

In the diagram ABC is an equilateral triangle.
D is the midpoint of AC.

∠ABD = ________°

AD = ________

BD = ________ (use the Pythagorean Theorem)

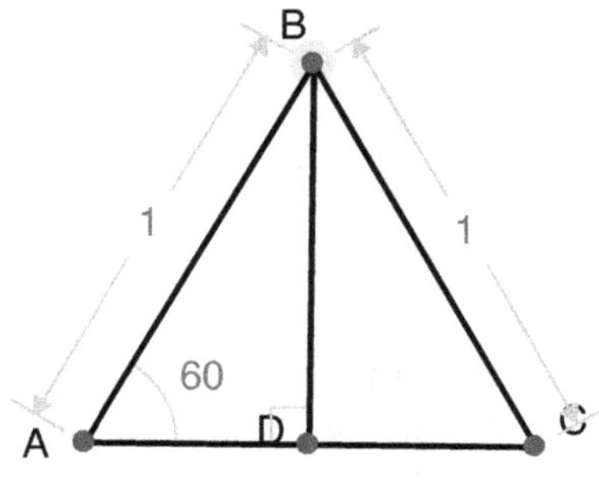

LAB #29 Right triangles - 30° and 45°

In this activity students:

- Use a standard triangle to find lengths as radicals and
- Halve an equilateral triangle to justify the $\frac{\sqrt{3}}{2}$ ratio

Ensure the software is set to measure angles in degrees.
The table refers to angles that are multiples of 15 degrees, all of which give radical expressions. The 50° is added to show not all angles have sides expressed as radicals. You may choose to omit this value and discuss this point at another time.

Solution

Angle (A)	Leg 1 (BC)	Leg 2 (AC)
15	$\sqrt{\frac{1}{2}-\frac{\sqrt{3}}{4}}$	$\sqrt{\frac{1}{2}+\frac{\sqrt{3}}{4}}$
30	$\frac{1}{2}$	$\frac{\sqrt{3}}{2}$
45	$\frac{\sqrt{2}}{2}$	$\frac{\sqrt{2}}{2}$
60	$\frac{\sqrt{3}}{2}$	$\frac{1}{2}$
75	$\sqrt{\frac{1}{2}+\frac{\sqrt{3}}{4}}$	$\sqrt{\frac{1}{2}-\frac{\sqrt{3}}{4}}$
50		

Justification for $\frac{\sqrt{3}}{2}$ requires students to use the Pythagorean theorem.

$$x^2 + 0.5^2 = 1$$

$$x^2 = \frac{3}{4}$$

$$x = \frac{\sqrt{3}}{2}$$

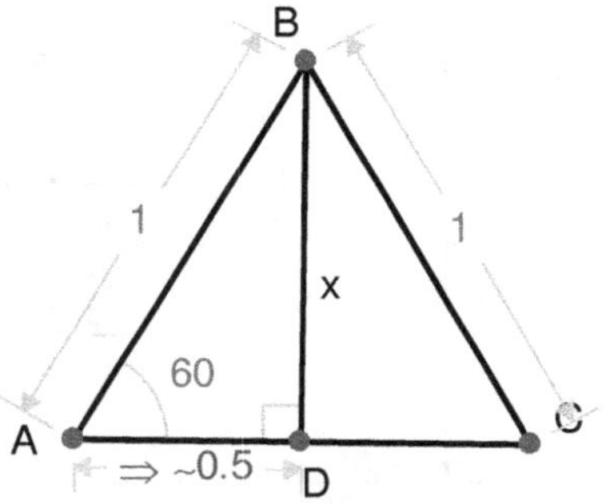

This applies to both the 30° and 60°. This may require a whole group demonstration and discussion. The 15° and 75° results can be obtained by using the addition formulas. E.g.

$$\sin 15° = \sin(45° - 30°)$$

$$= \sin 45° \cos 30° - \sin 30° \cos 45°$$

$$= \frac{\sqrt{2}}{2} \times \frac{\sqrt{3}}{2} - \frac{1}{2} \times \frac{\sqrt{2}}{2}$$

$$= \frac{\sqrt{6} - \sqrt{2}}{4}$$

GX gives

$$\Rightarrow \sqrt{\frac{1}{2} - \frac{\sqrt{3}}{4}}$$

Advanced students can show the equivalence to the expression given by GX. A calculator can readily verify the equivalence too.

$\angle$ABD = 30° , AD = 0.5, BD = $\frac{\sqrt{3}}{2}$

LAB # 30 Measures of "standard" triangles

Aim: Measure sides and ratios of right triangles inscribed in a unit circle.

Step 1 – Construct a unit circle

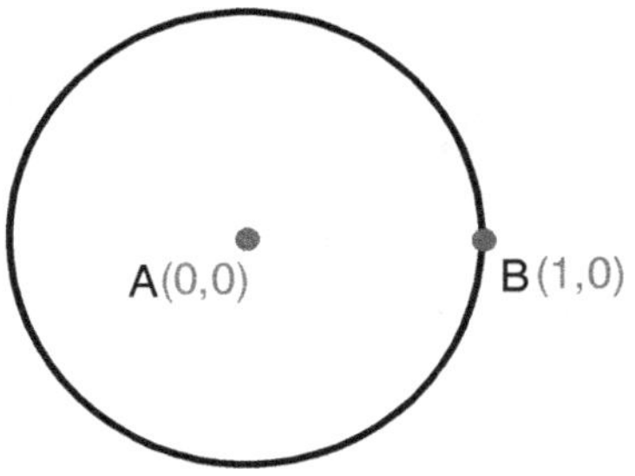

Construct a unit circle, center the origin.
- Construct point
- Constrain coordinates to (0, 0)
- Draw circle
- Constrain coordinates of point on circle to (1,0)

The circle is now of radius one unit.

Step 2 – Inscribe right triangle

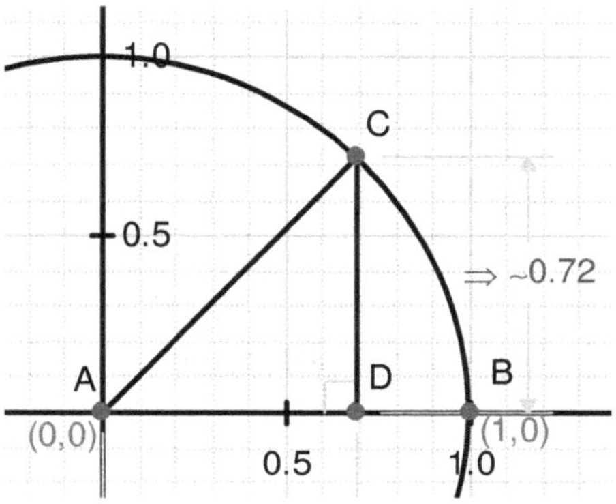

Show the grid

Plot another point on the circle in the first quadrant.

Select this point and the x-axis.

Construct | Perpendicular

Plot the intersection of the perpendicular with the x-axis.

Hide the perpendicular

Draw line segments to complete the triangle.

Drag the point (P) around and make sure that the triangle remains a right triangle.

Step 3 – Measure the triangle

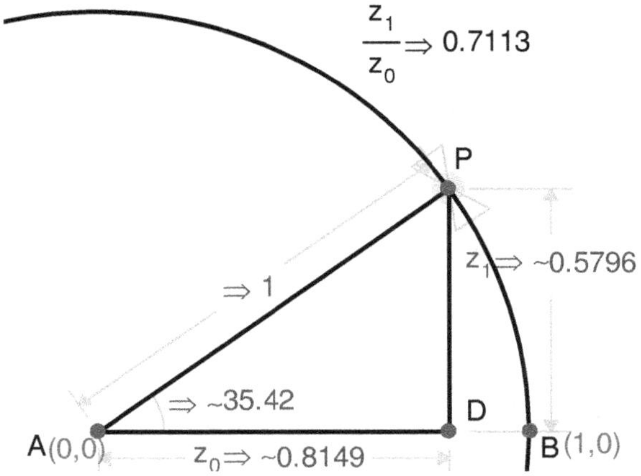

Check that Show Name = True:
- Edit | Settings | Math | Output

Measure the sides of the triangle and the angle the hypotenuse makes with the positive x-axis. ($\angle$PAD on the diagram)

Calculate the ratio of the y-coordinate to the x-coordinate.

- Draw | Expression
- Enter name of y-coordinate / name of x-coordinate, *e.g.* z[1]/z[0], to have GX do the calculation for you.

Step 4 – Fill in the table.

Use your GX drawing. Drag P to get as close as you can to the given measure.

Angle	x-coordinate	y-coordinate	hypotenuse	$\frac{y\text{-coordinate}}{x\text{-coordinate}}$
22°				
30°				
45°				
60°				
	0.3			

What is another possible answer for the last angle? ______________________________

LAB #30 Measures of "standard" triangles

In this activity students:

- Construct a unit circle centered on the origin with a right triangle inscribed in it;
- Measure x and y coordinates;
- Calculate the ratio of y coordinate to x-coordinate and
- Make measurements to build a table of values

The term "standard" triangle is being used to refer to right triangles with unit hypotenuse. Similarity arguments can then be employed to solve problems using these triangles. The specific values are used in Lab #31. In effect trigonometric tables are being built.

Trigonometric tables were used for all trigonometric calculations prior to the advent of electronic calculating devices. I talk to students about trig tables being derived from measurement of real triangles. The triangles had to be done very carefully and accurately and so we no longer need to do this. A picture of trig tables can be shown for interest value.

The table has only a few values in it. More can be done and a possible extension would be to create a table of x-co-ordinates from 1 to 90°. This may appeal to some students. The particular angles measured in this lab are used in the following activity where the triangles are scaled up to solve a problem derived from a practical situation. The aim is to reinforce the measurement concept before using the SOHCAHTOA algorithm.

A nice feature of this drawing is the way the triangle redraws itself outside the first quadrant. This provides a bridge to re-conceptualizing sine and cosine as circular functions.

Solution

Angle	x-coordinate	y-coordinate	hypotenuse	$\frac{y\text{-coordinate}}{x\text{-coordinate}}$
22°	0.927	0.375	1	0.404
30°	0.866	0.5	1	0.577
45°	.707	0.707	1	1
60°	0.5	0.866	1	1.73
73.3° or (360 – 73.3)°	0.3	1	1	3.33

LAB # 31 *Use of similar triangles to solve right triangles*

Aim: To solve right triangles using similarity.

Find the unknown in each situation. Use similar triangles and the table of values from Lab #30.

a. What height must be climbed to the mountain top?

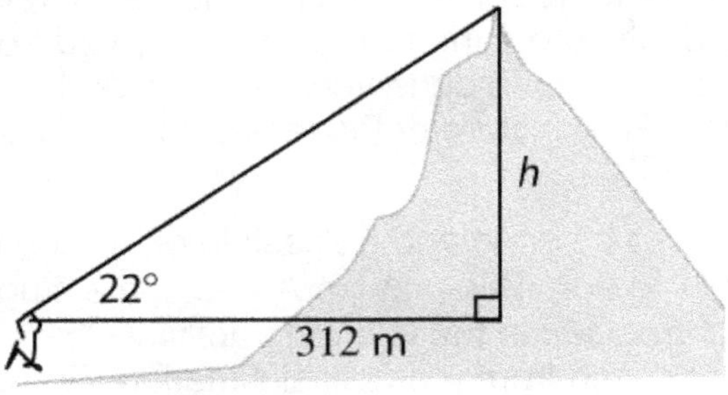

b. The corners are to be cut off a square panel to make an octagonal table. How long is the cut?

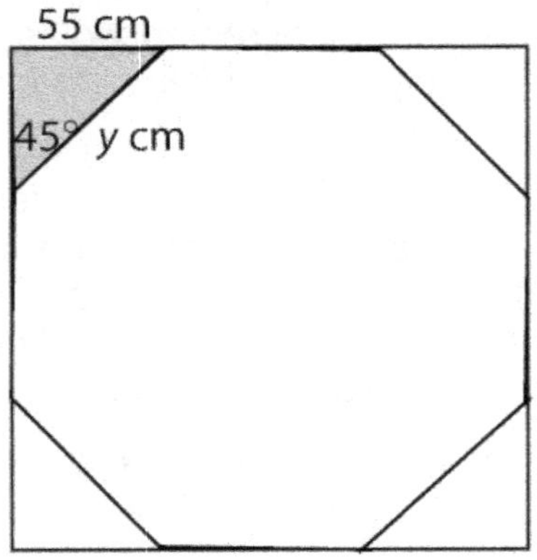

c. What angle does the 2 meter ladder make with the ground?

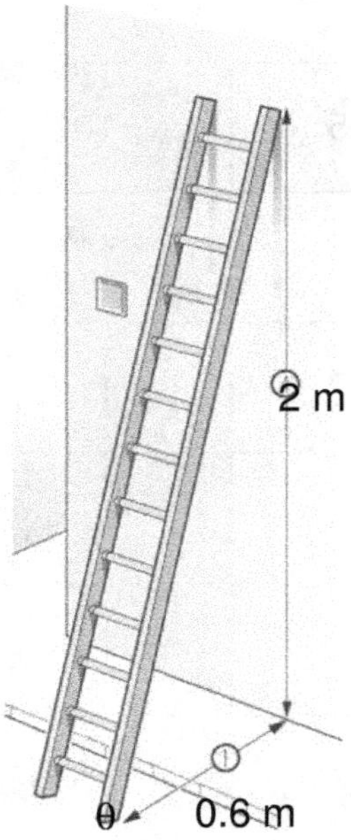

d. Going from A to X.
How far East and how far South is it?

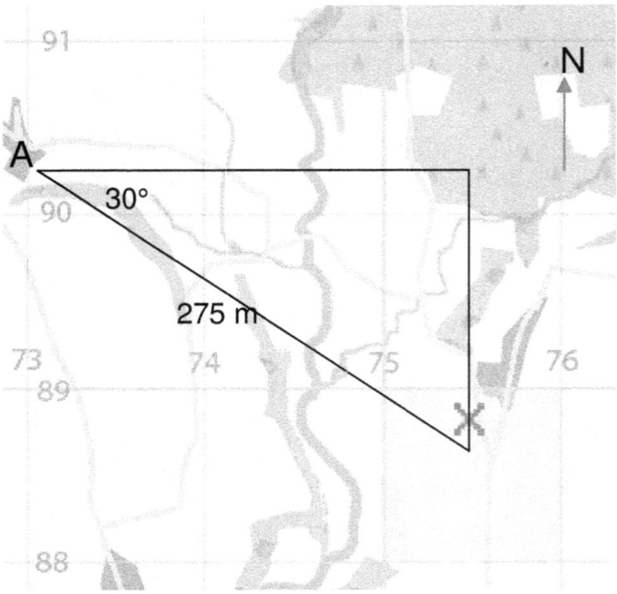

LAB #31 Use of similar triangles to solve right triangles

In this activity students

- Use the table of values from the previous lab to solve four problems.

This is an extension of the previous Lab and does not use GX. It is best done in conjunction with Lab #30.

By solving several problems by dilation of a "standard" triangle students can strengthen or develop concepts relating to the meaning of the trigonometric ratios. Note there has been no mention in this chapter of sine and cosine thus far. You may wish to make this connection explicit. The reasoning behind this relates to the underlying concept. While sine and cosine developed historically through measurement, within Trigonometry courses students are expected to work with a function concept for sine and cosine. The unit circle has been placed around the right triangle in Lab #30 and can lead into the function concept. Introducing sine and cosine at this later point may assist students to work with sine and cosine as functions rather than ratios within a right triangle.

Solution

a. $\frac{h}{312} = \frac{0.375}{0.927}$; h = 126 m

b. $\frac{y}{55} = \frac{1}{0.707}$; y =77.8 cm

c. $\frac{2}{0.6} = 3.33$; angle = 73.3°

d. $\frac{East}{275} = \frac{0.866}{1}$; East = 238 m

$\frac{South}{275} = \frac{0.5}{1}$; South = 137.5 m

LAB # 32 *Target practice*

Aim: To convert between Cartesian and polar coordinates.

A survey map has a grid superimposed on the map. Locations are given by grid references, either 6 digit or 8 digit for greater accuracy.

On the adjacent map,

- the grid marks squares 100 m by 100 m.
- References are always read from the Southwest corner or bottom left.

Distance between two grid references is calculated using the Pythagorean theorem.

Bearings are calculated using trigonometry.

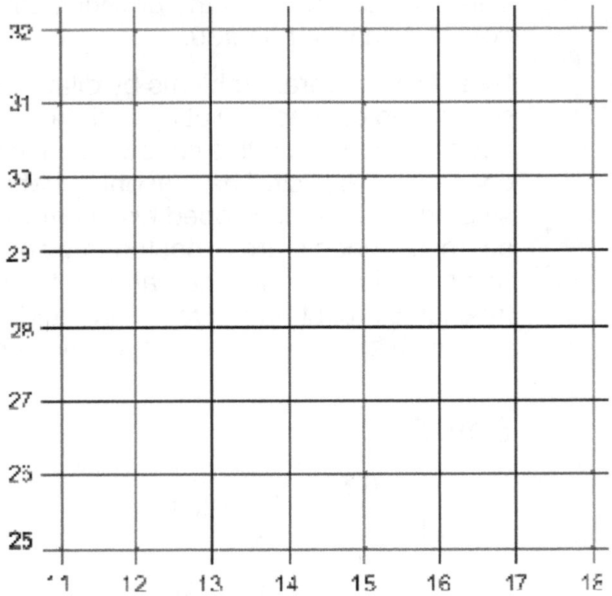

Example: Determine the distance and bearing of 126308 from 141265. Mark the points on the grid.

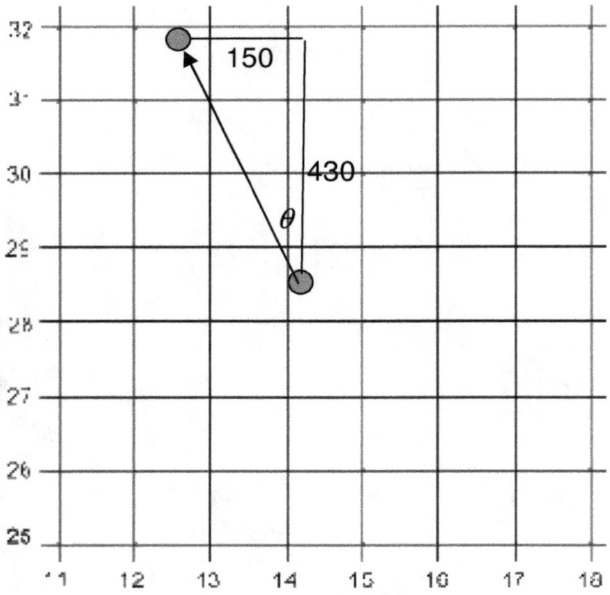

The change in horizontal (East-West) distance is 141 – 126 = 15 *i.e.* 150 m

The change in North-South distance is 308 – 265 = 43 *i.e.* 430 m

The distance between the points is $\sqrt{150^2 + 430^2} = 455\,\text{m}$ (use your calculator)

$$\theta = \tan^{-1}\left(\frac{15}{43}\right) = 19.2°$$

Bearings are measured clockwise from North.

Bearing is 360 – 19.2 = 340.8°T

Complete the table and plot on the grid above.

Grid reference		East-West distance	North-South distance	Distance	Bearing °T
From	To				
141265	126308	150	430	455	341
179312	144270				
110251	179319				
136311	167250				

Use *GX* to check your answers.

Draw a line segment
Constrain the endpoints to coordinates of the grid reference *e.g.* 1.41 for 141
Calculate | Distance of the line segment
Draw a vertical line

Constrain | Direction  to 90°
Select line segment and vertical line
Calculate angle.

LAB #32 Target practice

In this activity students:

- Work with map references;
- Calculate distance and bearing;
- Use right triangle trigonometry and
- Check answers using GX

Discuss with the whole group how grid references work and why distance and direction are required for firing at a target. Now GPS is likely to be used but prior to this military action would have involved scouts determining the grid reference of the target. The gunners would then calculate the direction and distance to the target and then set this up on the gun.

Solution

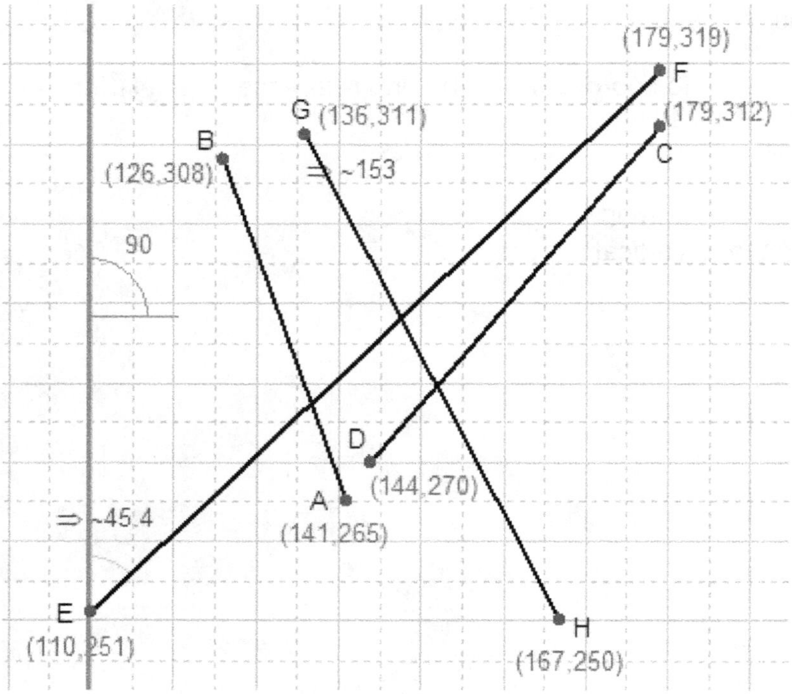

Grid reference From	To	East-West distance	North-South distance	Distance	Bearing °T
141265	126308	150 m	430 m	455 m	341°T
179312	144270	350 m	420 m	547 m	220°T
110251	179319	690 m	680 m	969 m	45.4°T
136311	167250	310 m	610 m	684 m	153°T

LAB # 33 *Sine formula*

Aim: Verify the sine formula

Create the drawing.

Draw a triangle

Check that Show Names is True.
  Edit | Preferences | Math | Output

Measure the length of each side
  Calculate | Real | Distance

Measure the size of each angle
  Calculate | Real | Angle

Calculate the ratio of the sine of each angle to the length of the opposite side.

e.g. $\frac{\sin A}{BC}$

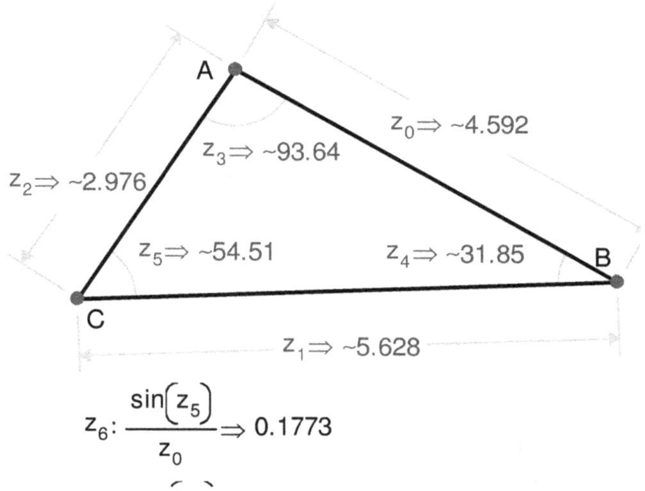

Drag the figure around.

What do you notice about the ratios?____

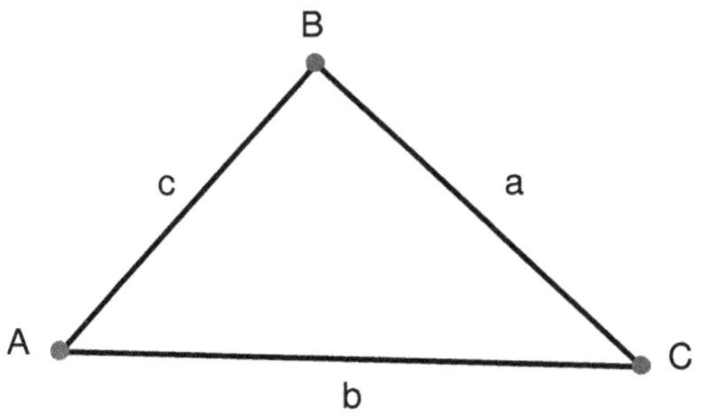

Complete the statement of the Sine formula based on your observations.

$$\frac{\sin A}{a} = \frac{\sin \quad}{\ldots\ldots} = \frac{\sin \quad}{\ldots\ldots}$$

LAB #33 Sine formula

In this activity students:

- Draw a triangle;
- Measure all sides and angles;
- Calculate the ratio and
- Verify the three ratios are equal.

Once students have the drawing they drag around vertices on their triangle and observe that the ratio for the sine of an angle to the length of the opposite side is the same for each angle. This verifies the Sine formula or law of sines.

The most common proof relies on calculating the triangle area using $\text{Area} = \frac{1}{2}ab\sin C$.

Calculating the area using a different angle each time gives

$$\frac{1}{2}ab\sin C = \frac{1}{2}bc\sin A = \frac{1}{2}ac\sin B$$

Divide by $\frac{1}{2}abc$ to produces one form of the Sine formula

$$\frac{\sin C}{c} = \frac{\sin A}{a} = \frac{\sin B}{b}.$$

Solution

The ratios are always equal.

$$\frac{\sin C}{c} = \frac{\sin A}{a} = \frac{\sin B}{b}$$

LAB # 34 *Law of cosines*

Aim: Explore the law of cosines

Part 1 – Calculate the third side

Create the drawing.

Draw a triangle

Constrain side BC to a

Constrain side AC to b

Constrain ∠C to C

Calculate | Distance AB

AB^2 = ________________

This is known as the Law of cosines or cosine formula.

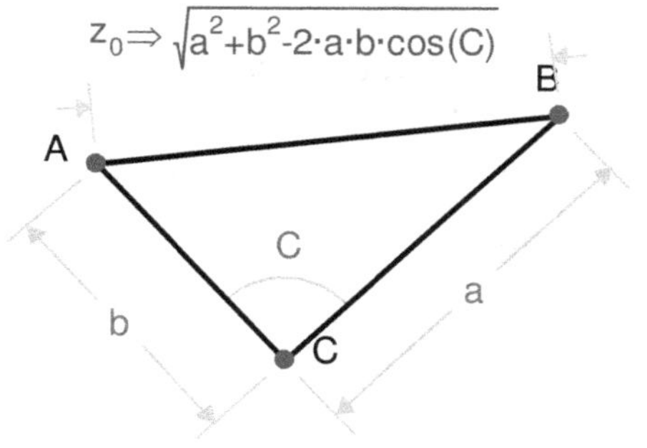

Repeat constraining a different pair of sides and the angle between them. For example.

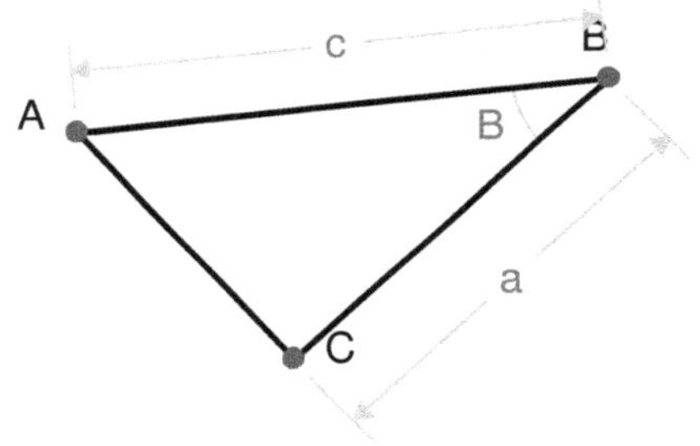

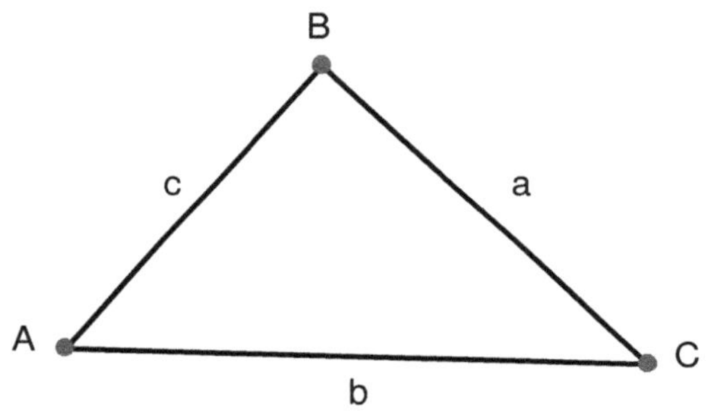

Use labeling consistent with the diagram on the right

Calculate | Symbolic | Distance for the third side.

Repeat for the third angle and pair of sides.

Use your results to complete the equations:

a^2 = ____

b^2 = ____

c^2 = ____

Part 2 – calculating the angle.

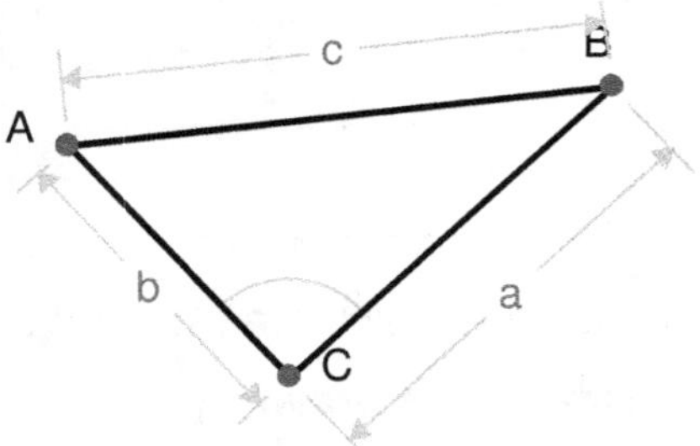

$$z_0 \Rightarrow \arccos\left(\frac{a^2+b^2-c^2}{2\cdot a\cdot b}\right)$$

Delete the constraints on all the angles. Constrain the three sides of the triangle as shown in the diagram.

Measure each angle
Calculate | Symbolic | Angle

Record the cosine of each angle

cos A = ______________________________

cos B = ______________________________

cos C = ______________________________

Derive one of these expressions from your formula in Part 1.

Note:

If cos$B$ = 0.6 then B = arc cos (0.6).

arccos() is the inverse of cos().

On a calculator it is often written as $\cos^{-1}$ or inv cos.

LAB #34 Law of cosines

In this activity students

- Use GX to calculate the length of the third side of a triangle from the other sides and the included angle;
- Calculate the angle from three sides and
- Derive the angle formula from Part 1.

A coordinate geometry proof is Lab #40.

In Part 1 of the Lab students use GX to generate a formula for the length of each side in terms of the other two sides and included angle according to the standard labeling of a triangle. It is nice the way GX generates the expressions as the law of cosines. This aims to emphasize the different substitutions or labelings possible.

In Part 2 the same approach is used to calculate each angle from the sides.

Part 3 is a pen and paper exercise to derive the Part 2 formula from the results in Part 1. This rearrangement often causes problems for students and extra support may be required for this step. One of the difficulties may be adding $2bc$ rather than divide by -$2bc$.

Solution

Part 1

$AB^2 = a^2 + b^2 - 2ab\cos C$

$a^2 = b^2 + c^2 - 2bc\cos A$

$b^2 = a^2 + c^2 - 2ac\cos B$

$c^2 = a^2 + b^2 - 2ab\cos C$

Part 2

$$\cos A = \frac{b^2 + c^2 - a^2}{2bc}$$

$$\cos B = \frac{a^2 + c^2 - b^2}{2ac}$$

$$\cos C = \frac{a^2 + b^2 - c^2}{2ab}$$

Derivation:

$$a^2 = b^2 + c^2 - 2bc\cos A$$

$$a^2 - b^2 - c^2 = -2bc\cos A$$

$$\frac{a^2 - b^2 - c^2}{-2bc} = \cos A$$

$$\frac{b^2 + c^2 - a^2}{2bc} = \cos A$$

LAB # 35 *Solving non-right triangles*

Aim: To use Geometry Expressions to check solutions of triangles.

Solve each problem.

It is likely that your teacher will want you to be able to solve these problems using pencil and paper. GX can help you to visualize the problem and build your skills.

1. What are the measures of the angles in a 6 – 7 - 10 triangle?
2. In triangle ABC, AC = 34.5 m, BC = 49.2 m and ∠B = 41.79°. Find the two possible values for angle CDB.

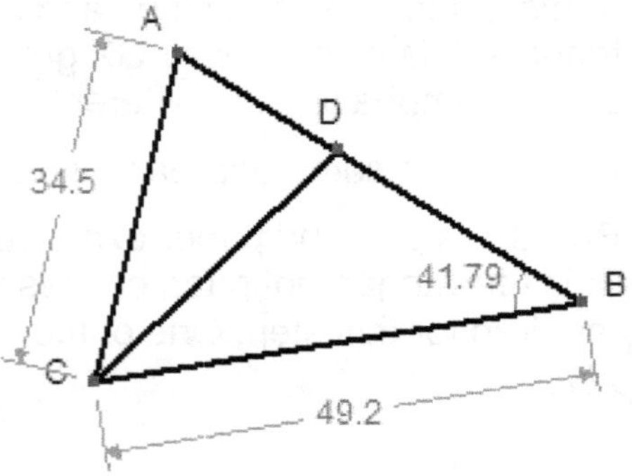

3. A ship leaves port at 2:00 am and sails in a direction of 120°T at a speed of 14 knots. Another ship leaves the port at 2:30 am and sails in a direction of 45° T at a speed of 17 knots. How far apart are they at 5 am?

   Note:

   120°T means 120 degrees measured clockwise from due North. It is the same as 30° South of east.
   45° T is the same as 45 East of North or NE.
   One knot is one nautical mile per hour.

   Extension: What is the bearing of the second boat from the first boat at 5 am?

4. Express θ (∠ACD) in terms of x.

   ∠ABC = 90°
   AD = 5, BD = 2, BC = x

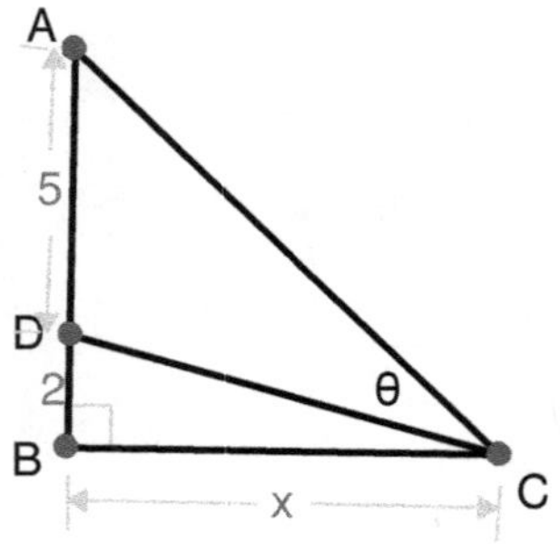

LAB #35 Solving non-right triangles

In this activity students:

- Solve some problems using the sine formula and the law of cosines and
- Use *Constraints* in GX to draw the diagram and check the calculation.

Where the triangles are fully specified you can set the *Constraints* and calculate using GX.

This is best done after the theory has been taught. You may like to set a problem for homework before using GX. The drawings assist students to visualize the problem to scale and this can strengthen their ability to check their answers for reasonableness.

Alternatively GX could be used to do the calculations.

Solution

1.

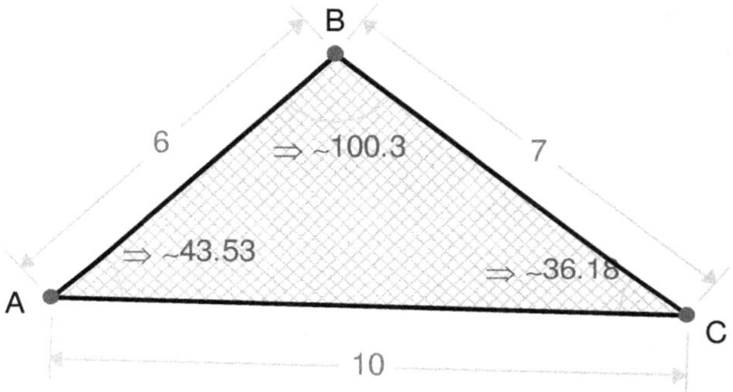

2.

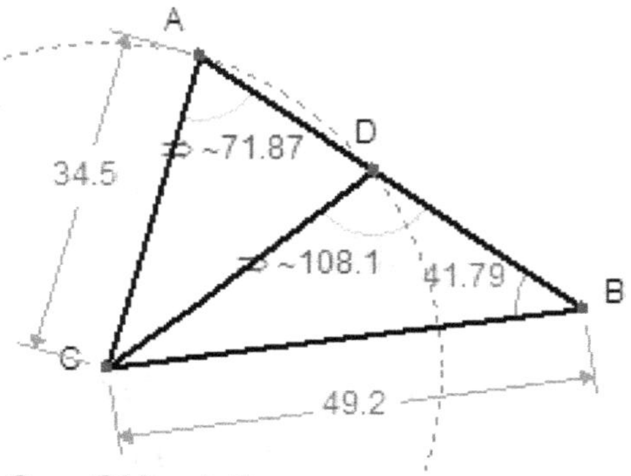

One GX solution:

Draw triangle ABC
Constrain BC to 49.2
Constrain ∠ABC to 41.79°
Draw Circle center A radius AC
Constrain radius to 34.5
Construct Intersections of AB and circle

3.

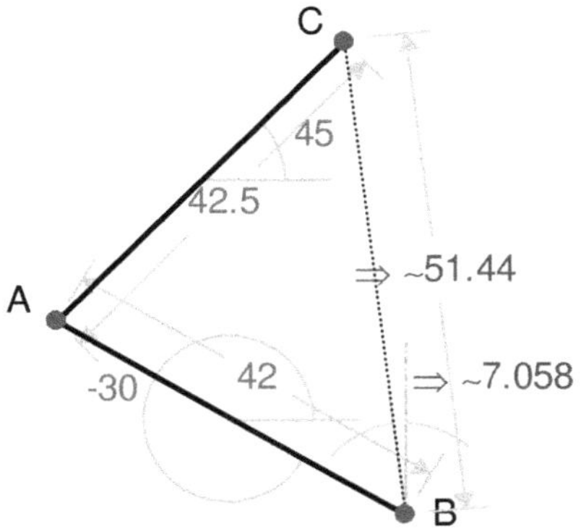

Extension: direction of BC is 360 – 7.06°T

4.

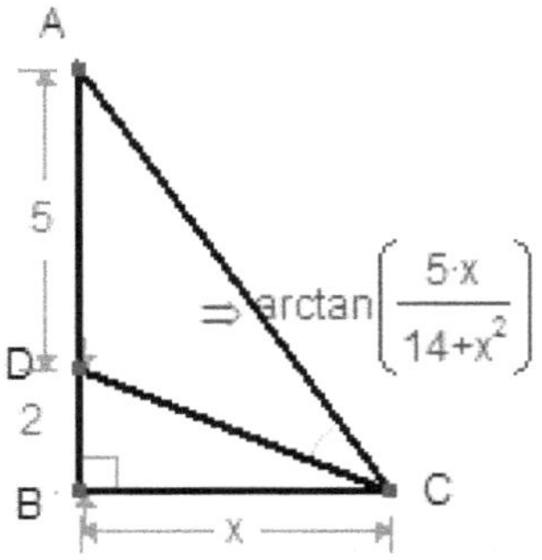

Alternative using law of cosines,

$$\cos\theta = \frac{(4+x^2)+(49+x^2)-25}{2\sqrt{4+x^2}\sqrt{49+x^2}}$$

LAB # 36 *How much higher?*

Aim: To use Geometry Expressions to check solutions of triangles.

Tom and Maria love walking and they are climbing a mountain. Tom wonders how much higher they have to climb to the top. Tom measures the angle of elevation to the top of the mountain at 33°. 400 m closer to the mountainTom measures the angle of elevation again at 39°.

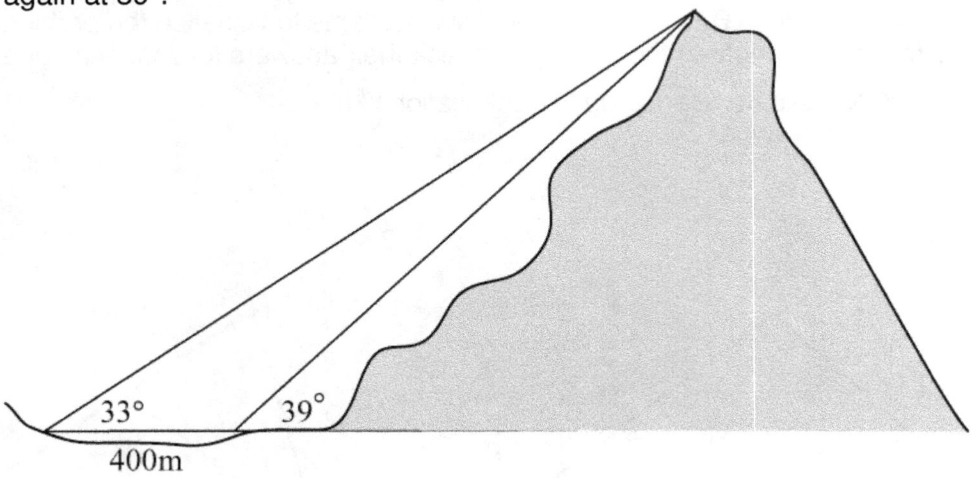

How high is the mountain (relative to Tom's current position)? ____________ m

Construct the figure in GX.

Make the appropriate measurements to answer the question.

Check your answer using the Sine Formula with pen and paper.

EXTENSION

Write a similar problem.

Solve your problem using GX and using pen and paper.

Trade your problem with a partner and solve each others problems.

LAB #36 How much higher?

In this activity students:

- Use the sine formula to solve a problem and
- Check their solution using GX

You may like to set the problem for HW or in class time before going to the computers and using GX. The drawing should be straight forward.

Solution

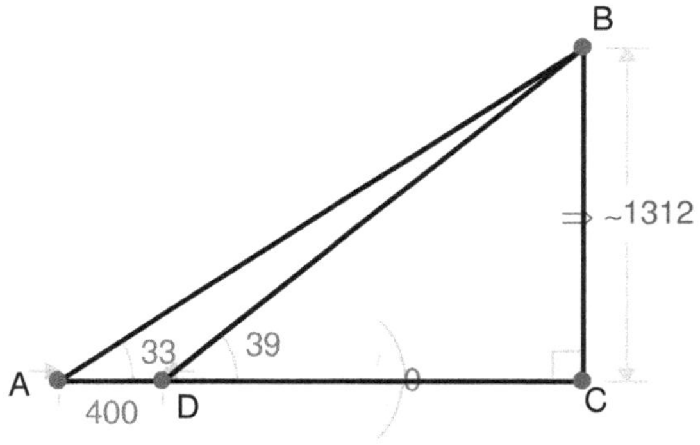

$\angle ABD = 39 - 33 = 6°$ Exterior Angle Theorem

$$\frac{\sin \angle ABD}{400} = \frac{\sin 33}{BD}$$ Sine formula

$$BD = \frac{400 \sin 33}{\sin 6} = 2084$$

$$\sin 39 = \frac{BC}{BD}$$

BC = BD sin 39° =1312 m

LAB # 37 *Stranded on an Island*

Aim: To use Geometry Expressions to check solutions of triangles.

Imagine that we are stranded on an island (the playground).

We have tape measures and protractors to make measurements.
What can we measure off the island without leaving the island?

One group of students wanted to measure the width of the basketball court across the road (sea).

They recorded their results as sketches.

AB is the baseline (on the "island" – students measured this inside the playground)

CD is one edge of the court (off the "island" – the distance to be calculated)
Figure 1 is a plan showing the baseline AB on the island and the edge of the court CD off the island.

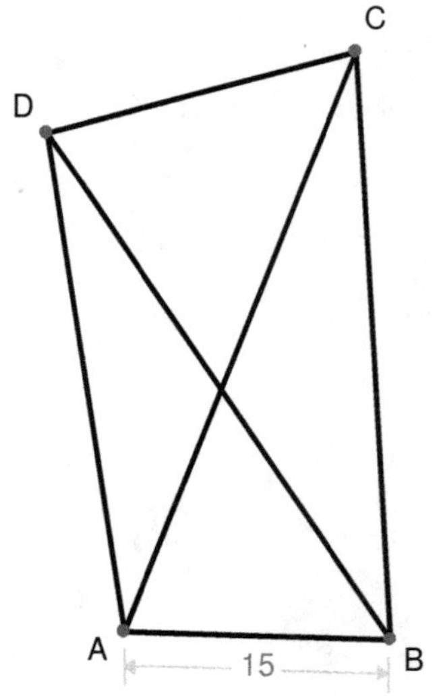

Figure 1

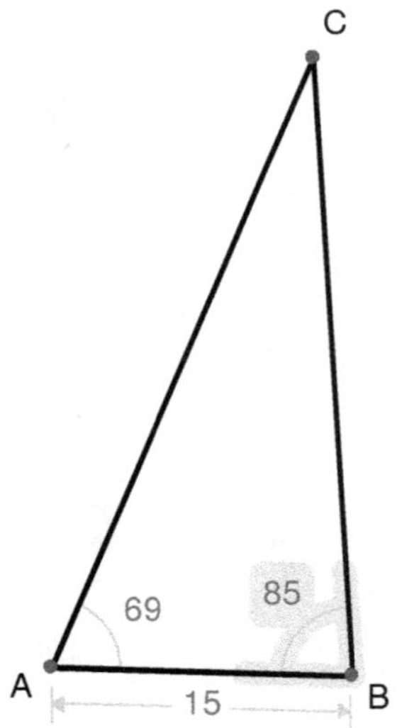

Figure 2

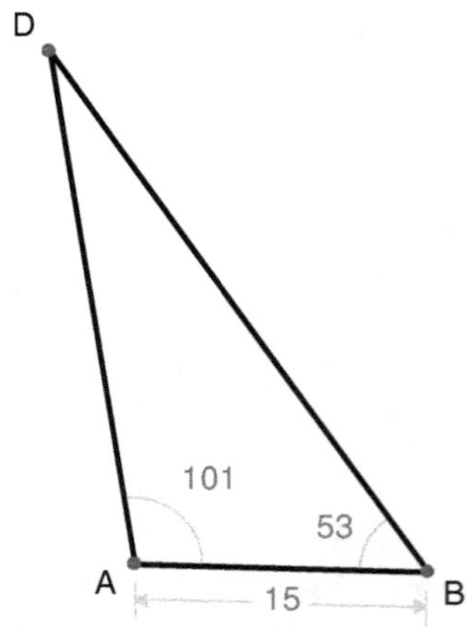

Figure 3

Solve the problem using the sine and cosine rules on paper.

In Figure 2 ∠ACB = _____

Calculate AC using the sine rule

In Figure 3 ∠ADB = _____

Calculate AD using the sine rule

In Figure 1 ∠DBC = _____

Calculate CD using the cosine rule

Draw the sketch in GX.
Set the *Constraints* for the measurements
Fill in the table of measures

AC	
AD	
∠DAC	
CD	

LAB #37 Stranded on an Island

In this activity students:

- Solve a complex multi-triangle problem using the Sine formula and law of cosines and
- Check the solution using GX.

This is a great activity to do practically and then use the measured values instead of those provided in the Lab. A colleague set up a series of activities for students to do which he called *measuring the inaccessible*. He took pictures around the school and placed word art on the digital image to identify the required measurement. It was up to the student groups to work out what they would measure and the method they would use.

Equipment:

- Measuring tape and/or trundle wheel
- Device for measuring angles in the horizontal plane. There are some relatively cheap devices available for the home builder. At the site where the photos were taken we used large protractors. These were made by the students, had a diameter of 20" and were acceptably accurate to one degree.

The diagrams can be difficult to follow. If you do not go outside you can model the situation on the classroom floor. Adding the labels consistent with the lab will also help. This will assist students to appreciate what they are calculating. Note: the labeled points, such as A, are the same point in each Figure.

Solution

$\angle ACB = 26°$

$AC = 34.1$

$\angle ADB = 26°$

$AD = 27.3$

$\angle DBC = 32°$

$CD = 18.1$

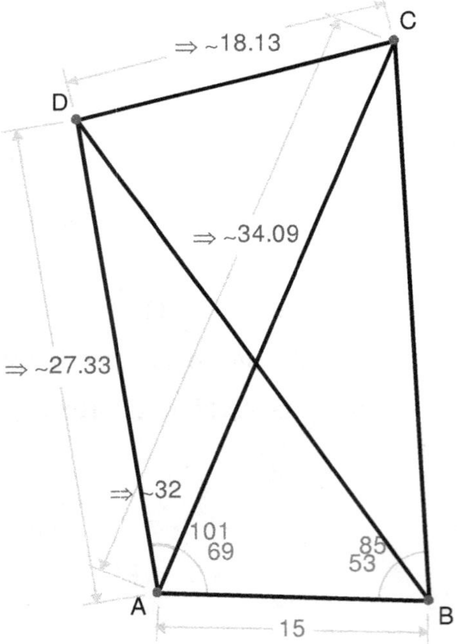

Chapter 9 – Coordinate Geometry

Teacher Notes

Coordinate systems can be used to prove many results in geometry. Being aware of where coordinate geometry provides a convenient or productive approach is beneficial.

This chapter uses GX to assist in developing coordinate geometry arguments. Constraining points is consistent with *without loss of generality* arguments. Students are able to follow the steps involved in the coordinate geometry proof by symbolically calculating the appropriate parts of the diagram.

The labs in this chapter often begin with numerical examples to develop understanding of the figure. Extending this to use *Symbolics* assists students to generalize and to develop proofs.

The initial labs are short and seek to prove elementary formulas. With more complex results using GX helps to focus on the key steps and ideas required in the proof. This is particularly useful for students who find themselves focused on the individual steps and unable to appreciate the big picture.

The labs use coordinate geometry arguments to prove

- Distance formula
- Midpoint formula
- The quadrilateral formed by connecting adjacent midpoints of a quadrilateral form a parallelogram
- The diagonals of a rhombus bisect each other at right angles
- The angle in a semicircle is a right angle
- The law of cosines
- The Subtraction formula for cos $(A - B)$

LAB # 38 *Distance formula*

Aim: To understand and apply the distance formula.

Part 1

What is the distance between the points (1, 5) and (7, –3)?

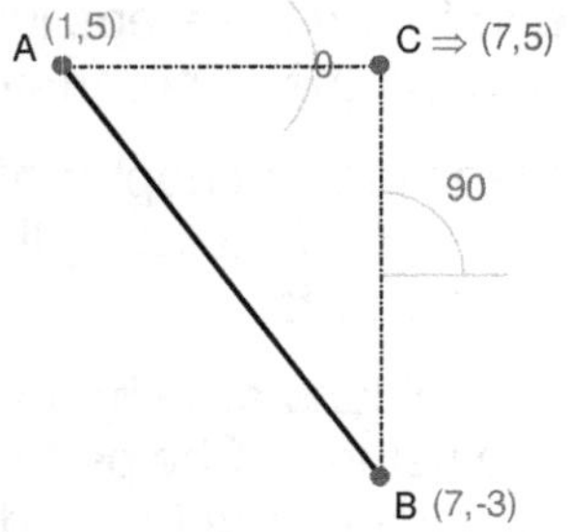

Draw line segment AB
Constrain end points
Constrain | Coordinates
Draw line segment AC
Constrain | Direction of AC to 0 (make the line horizontal)
Draw line segment BC
Constrain | Direction to 90° (make the line vertical)
Measure the coordinates of C
Calculate | Coordinates
Measure the lengths AC and BC.
Calculate | Distance

Use these values and the Pythagorean Theorem to calculate $(AB)^2$ ________________

Check your answer using GX.

Calculate the distance between (1, 5) and (–1, –3) using the Pythagorean Theorem.

$(AB)^2$ = ____

Check your answer using GX.
Alter the *Constraints* for B. Double click on the coordinates of B and enter the new values.

Part 2 The formula

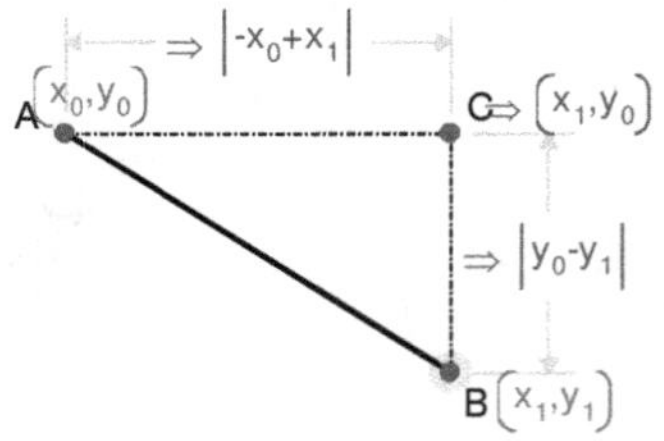

Generalize the *Constraints*
Delete constraints for A and B.
Constrain | Coordinates of A to the default value (x_0, y_0).
Constrain | Coordinates of B to the default value (x_1, y_1).

Calculate the length of AB
Select AB
Calculate | Length

AB = ____

Using the Pythagorean Theorem $(AB)^2$ = ______________________________________

LAB #38 Distance formula

In this activity students:

- Use GX to calculate the distance between two points numerically and symbolically and
- Derive the distance formula

Part 1 uses numerical examples. You may like to turn the axes and grid on,  , View | Axes, View | Grid to emphasize working with coordinates. The right triangle is included to emphasize the distance formula as an application of the Pythagorean Theorem. Note: in Part 2, the points A and B can be dragged around maintaining the right triangle.

In the drawing provided in the Lab, the line properties have been changed to show a dotted line. Select the line or other object, right click and choose Properties. The line style and color can be changed to improve the diagram. In this case the dotted line enables the viewer to focus on the line segment but show the right triangle as well.

Part 2 modifies the diagram by generalizing the coordinates and allowing the software to calculate the formula, the generalized case. The coordinates of the points are constrained to specific values and then this is generalized through the use of *Symbolics*.

You may wish to discuss the need for the absolute value in determining the side lengths of the right triangle. These appear in the diagram shown in Part 2.

Solution

Part 1

$(AB)^2 = 100$

Calculate the distance between (1, 5) and (-1, –3) using the Pythagorean Theorem.

$(AB)^2 = 68$

Part 2

$AB = \sqrt{(x_1 - x_0)^2 + (y_1 - y_0)^2}$

$AB^2 = (x_1 - x_0)^2 + (y_1 - y_0)^2$

LAB # 39 *Midpoint formula*

Aim: To develop and use the midpoint formula.

Draw line segment AB

 Construct | Midpoint of AB

Measure the lengths AC and BC.
Calculate | Real | Distance

Drag A and B around.

What do you observe about the lengths of AC and BC?

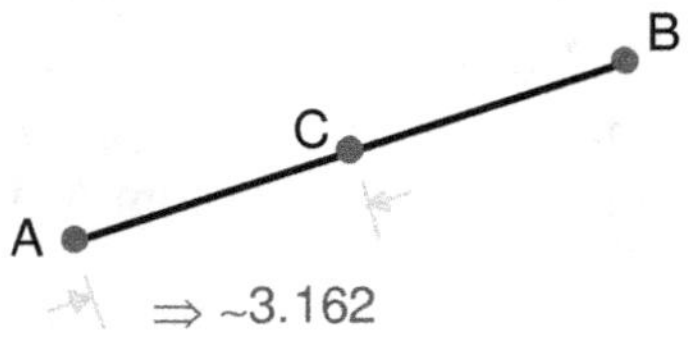

____

Delete or hide the lengths AC and BC.

 Constrain end points (4,2) and (10,4)

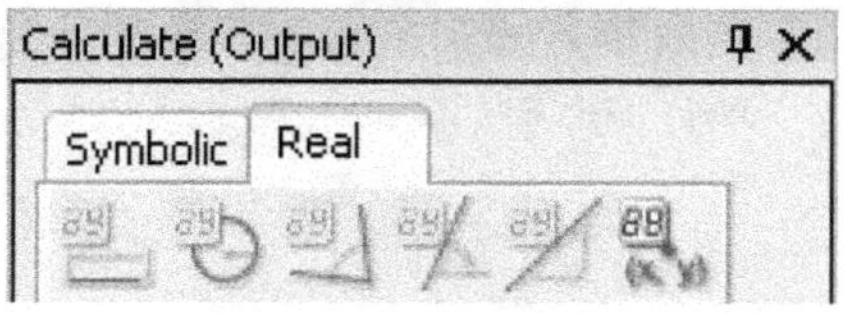

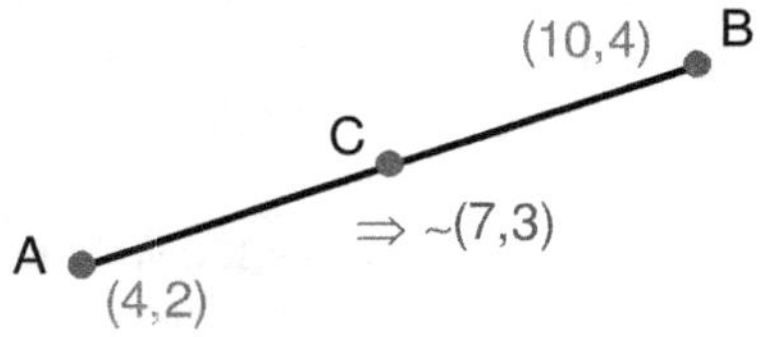

Measure the coordinates of the midpoint..
Change the end points.

How can the coordinates of the midpoint be calculated from the end points?
(Hint: Think about averages)

____

Delete the *Constraints* for the endpoints.

Constrain the end points to the default values or use other symbols.
Calculate | Coordinates (Symbolic)

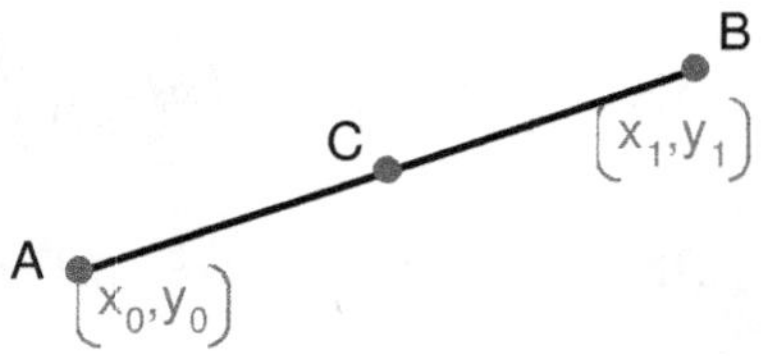

The coordinates of the midpoint are (__________ , __________)

Write a sentence explaining why. ____

____

____

LAB #39 Midpoint formula

In this activity students:

- Draw a line and construct it's midpoint
- Use *GX* to investigate the properties of the midpoint and derive the formula
- Check their formula using *GX's* symbolic output.

The lab starts with a numerical example and uses *GX* to help the student discover the general formula for the midpoints' coordinates.

Solution

AC and BC are equal in length.

The x coordinate of the midpoint is the average of the x coordinates and the y coordinate is the average of the y coordinates.

The midpoint is $\left(\frac{x_0 + x_1}{2}, \frac{y_0 + y_1}{2}\right)$. The coordinates are calculated by averaging the x and y coordinates.

LAB # 40 *Midpoints of a quadrilateral*

Aim: To develop a conjecture about the quadrilateral formed by connecting the midpoints of a quadrilateral.

Draw a quadrilateral

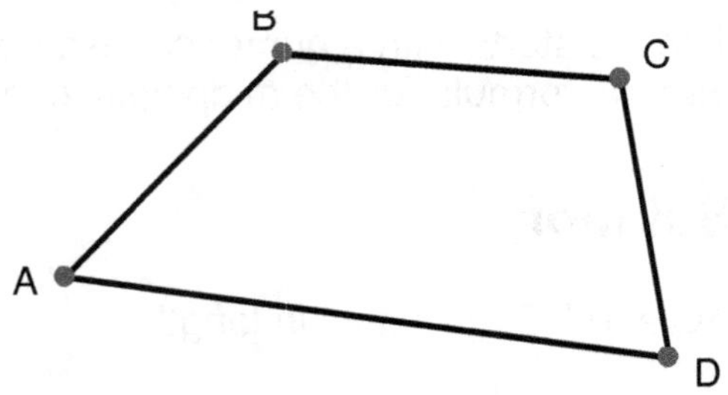

Construct midpoints of each side
Join adjacent midpoints to form quadrilateral EFGH.

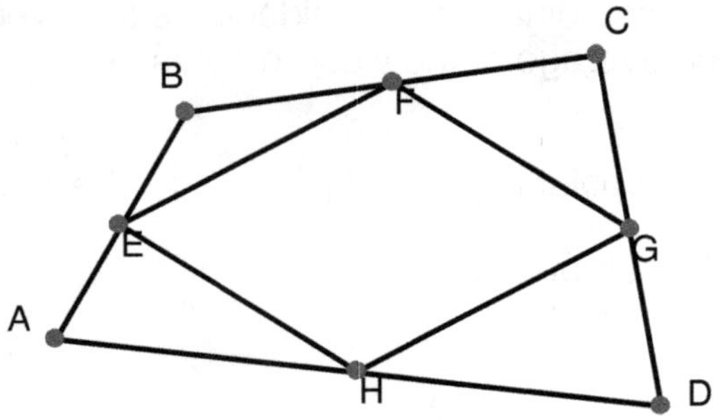

Measure the slope of each side of EFGH
    Select line segment
    Calculate | Slope.

What do you notice?

____

Drag the vertices to see if your conjecture still holds.

Complete the statement:
The quadrilateral formed by connecting adjacent midpoints of any quadrilateral is

____

Outline for a coordinate geometry proof;

- Assign coordinates to the vertices of the quadrilateral constrain the coordinates.
- Calculate the coordinates of the midpoints.
- Calculate the slope between adjacent midpoints.
- Compare the slopes of the opposite sides.

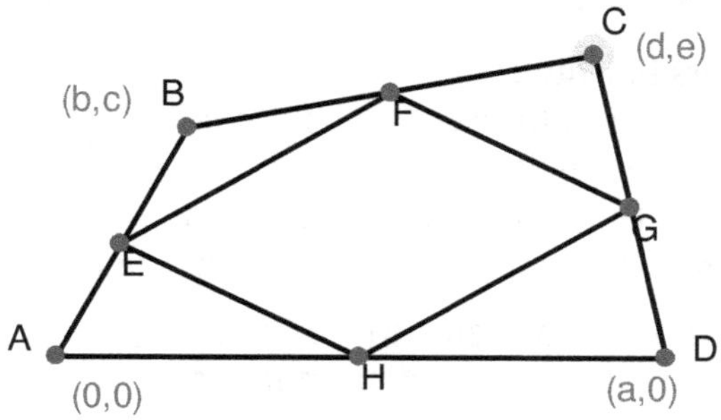

This diagram shows calculations done by GX.

Use GX to carry out the steps required to prove your conjecture.

Complete the proof below. (You can use the same variables and calculation results shown in the GX drawing)

Let the vertices of quadrilateral ABCD be (0, 0), (b, c), (d, e), $(a, 0)$
Let E be the midpoint of AB

The coordinates of E are (____________ , ____________)

Let F be the midpoint of BC

The coordinates of F are (____________ , ____________)

Let G be the midpoint of CD

The coordinates of G are (____________ , ____________)

Let H be the midpoint of AD

The coordinates of H are (____________ , ____________)

The slope of EF = ____________________________________

The slope of HG = ____________________________________

The slope of EH = ____________________________________

The slope of FG = ____________________________________

EF and HG are ____________________________________

FG and EH are ____________________________________

EFGH is a ____________________________________

LAB #40 Midpoints of a quadrilateral

In this activity students:

- Construct a parallelogram by joining the midpoints of a quadrilateral.
- Investigate properties of the parallelogram
- Use GX to do the calculations used in the proof and
- Prove the result using coordinate geometry methods.

The proof requires knowledge of:

- Calculating coordinates of midpoints
- Parallel lines have equal slopes
- Algebraic manipulation skills.

This lab begins with an exploration phase designed to encourage students to conjecture that opposite sides of the inscribed quadrilateral are parallel. It is important that students drag vertices to establish that their conjecture applies to many cases.

The proof is highly scaffolded. Each calculation can be performed on the drawing in GX.

Solution

The slopes of opposite sides of EFGH are equal.

The quadrilateral formed by connecting adjacent midpoints of any quadrilateral is a parallelogram.

Outline for proof:
Without loss of generality, let the vertices of quadrilateral ABCD be (0, 0), (b, c), (d, e), $(a, 0)$

Let E be the midpoint of AB. The coordinates of E are $\left(\frac{b}{2}, \frac{c}{2}\right)$

Let F be the midpoint of BC. The coordinates of F are $\left(\frac{b+d}{2}, \frac{c+e}{2}\right)$

Let G be the midpoint of CD. The coordinates of G are $\left(\frac{a+d}{2}, \frac{e}{2}\right)$

Let H be the midpoint of AD. The coordinates of H are $\left(\frac{a}{2}, 0\right)$

The slope of EF = $\frac{e}{d}$. The slope of HG =. $\frac{e}{d}$

The slope of EH = $\frac{c}{b-a}$. The slope of FG = $\frac{c}{b-a}$

EF || HG, FG || EH (equal slopes) therefore EFGH is a parallelogram.

LAB # 41 *Diagonals of a rhombus*

Aim: To discover and prove the relationship between the diagonals of a rhombus.

Exploration

Draw a rhombus.

Draw a quadrilateral.
Constrain the lengths of each side to *a*.

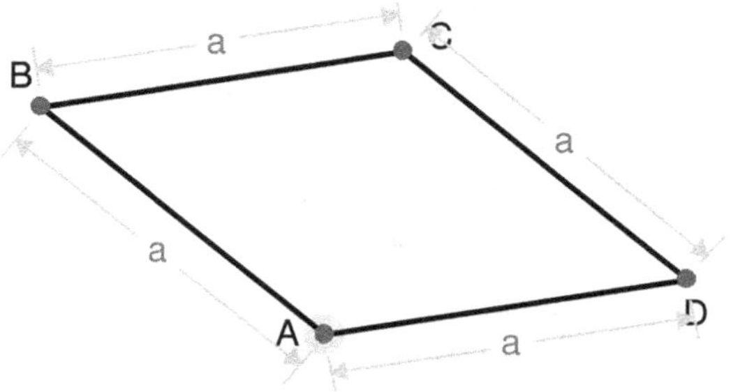

Draw the diagonals.

Measure the angle between the diagonals. The angle is __________ °

Draw a point at the intersection of the diagonals.
Measure the lengths from the intersection to each vertex.

Write a conclusion ____

____

Create the drawing using GX.

Make measurements and calculations on your drawing as indicated in the outline proof below.

Outline proof

Without loss of generality (and to make the algebra simpler) set the coordinates of A to (0, 0) and B to (a, 0), C to ($a + c$, b) and D to (c, b).

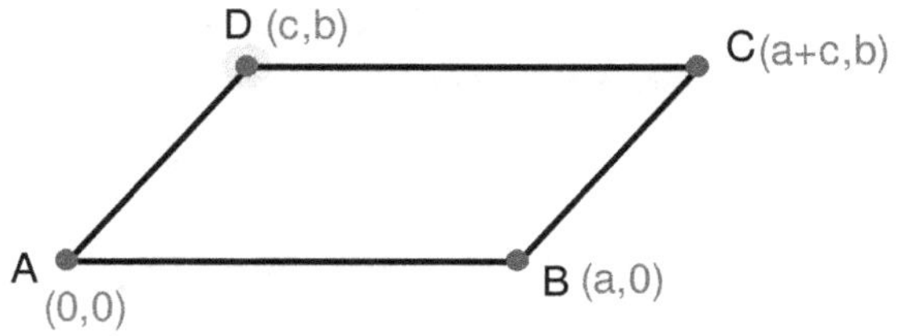

Notice that this doesn't look like a rhombus.

Diagonals of a parallelogram bisect each other.

Show the midpoints of the diagonals are coincident.

Perpendicularity

Calculate the slopes of AC and BD.

Multiply the slopes together.

Use facts about a rhombus i.e. The length AD must equal AB so

$a^2 = c^2 + b^2$

Simplify the expression.
(Note: If the product of the slopes is –1 the lines are perpendicular.)

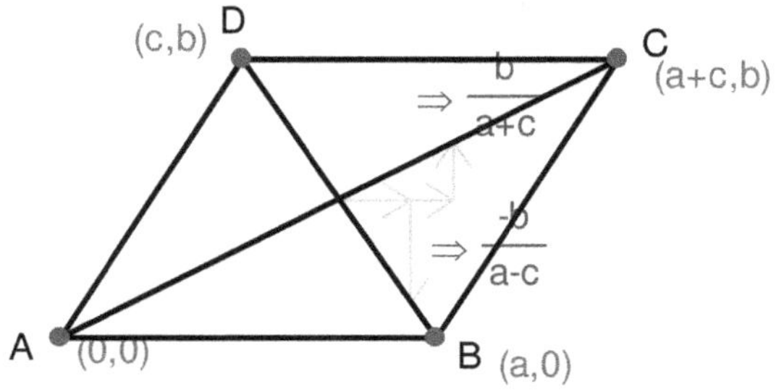

Write a coordinate geometry proof.

LAB #41 Diagonals of a rhombus

In this activity students:

- Draw a rhombus using *Constraints*;
- Discover that the diagonals are perpendicular and bisect each other and
- Prove the result using coordinate geometry methods.

The proof requires knowledge of:

- Slopes of perpendicular lines are negative reciprocals;
- The distance formula and
- Algebraic manipulation skills.

This lab begins with an exploration phase to familiarize students with the rhombus and its properties. The guided exploration leads to conjecturing the perpendicularity and bisection of the diagonals.

This result was proved using a different approach in Lab #12.

Extensions could include:

The converse, if the diagonals of a quadrilateral are perpendicular and bisect each other then the quadrilateral is a rhombus.

Constrain the diagonals to make a right angle

Draw the intersection point of the diagonals

Constrain | Distance from intersection to vertex to distance *a*.

Repeat from the intersection to the opposite vertex.

Similarly for other diagonal to create diagonals that bisect each other.

Calculate | Distance of the quadrilaterals sides.

Diagonals of a rhombus bisect the angles.

Solution

The angle between the diagonals is 90°

A sample proof is shown below. Note: No scaffolding of the proof is shown in this lab unlike the previous lab.

Midpoint of AC is $\left(\frac{a+c}{2},\frac{b}{2}\right)$ Midpoint of BD is $\left(\frac{a+c}{2},\frac{b}{2}\right)$

AC and BD have the same midpoint. Therefore AC and BD bisect each other.

Slope of AC = $\frac{b}{a+c}$, Slope of BD = $\frac{-b}{a-c}$

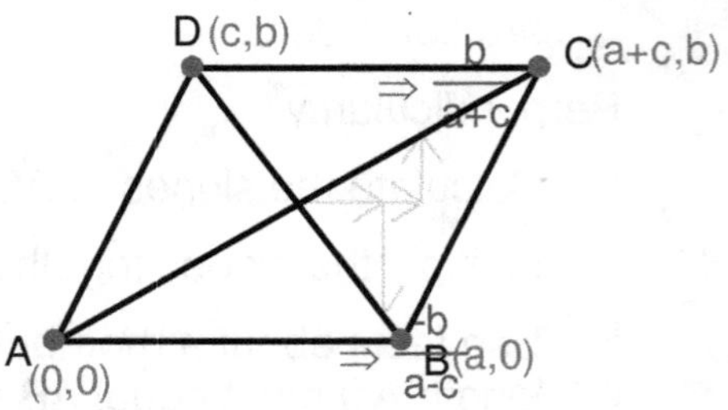

Slope of AC × slope of BD = $\frac{b}{a+c} \times \frac{-b}{a-c}$

$$= \frac{-b^2}{(a+c)(a-c)}$$

$$= \frac{-b^2}{a^2-c^2}$$

$= \frac{-b^2}{b^2}$, ($b^2 + c^2 = a^2$) as AD = AB

$$= -1$$

$\therefore$ AC $\perp$ BD

LAB # 42 *Angle in a semicircle*

Aim: To prove the angle in a semi-circle theorem.

STEP 1 - Exploration

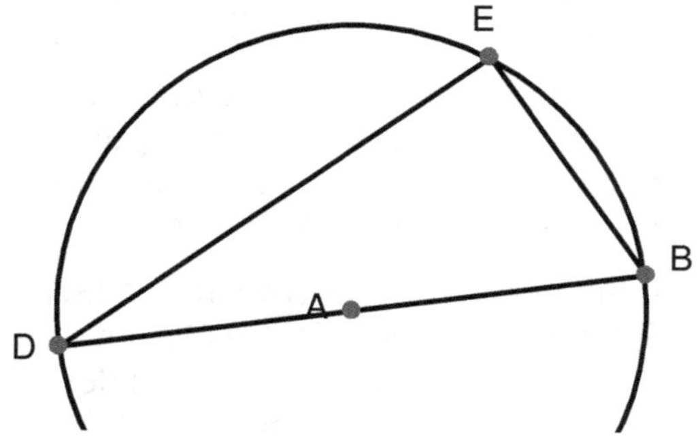

Construct the diagram.
- Draw a line segment
- Construct the midpoint
- Construct a circle
  - Click on the midpoint and drag out to the end of the line segment.
- Place another point on the circle
- Connect to end points of diameter

Measure the angle in the semi-circle.
Calculate | Real | Angle

The angle is ______ °

Drag points to see how the value changes.

STEP 2 - Proof

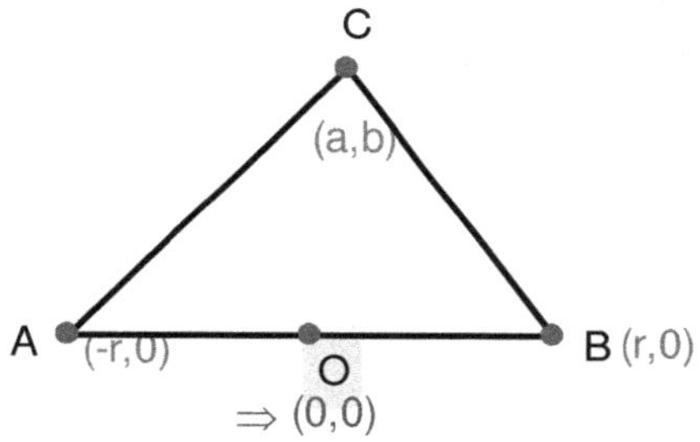

Create a new drawing using GX.

Draw triangle ABC

Constrain the coordinates of A to $(-r, 0)$, B to $(r, 0)$, C to (a, b) and O (0, 0).

Note:
- This is equivalent to AB being the diameter of the circle centre O and radius r.
- A without loss of generality argument is being used.
- OC = r but is not constrained in the diagram.

Use your drawing in GX to:
Calculate the slopes of AC and BC
(in terms of r, a and b)
Calculate | Symbolic | Slope
(make sure Show Names is true Edit | Preferences | Math | Output)

Multiply the slopes together.

Draw | Expression 

Write a coordinate geometry proof.

Note: the ideas needed in the proof are similar to the previous lab.

LAB #42 Angle in a semicircle

In this activity students:

- Construct the angle in a semi circle and measure the angle;
- Set up coordinates to develop proof;
- Use GX to calculate slopes and the product of slopes and
- Write out a proof.

This is similar to previous lab in terms of pre-requisite knowledge required and the use of a without loss of generality argument. The algebra is simpler as there is only the need to establish perpendicularity. You may prefer to do this activity before the diagonals of a rhombus (Lab #41) for this reason

If the angle is measured in Symbolic mode you get a terrible mess, make sure that the Real tab is chosen in the Output window. Then, I constructed the circle, created a diameter (used Constrain | Incident on a chord and the center), and finished the triangle. Calculate | Symbolic gave me 90. It is not possible in GX to constrain coordinates and point C to the circle. When a user attempts to do this a message appears informing the user that there is a conflict. This is because you are supplying more information than is needed to create the drawing, making it over-constrained. With duplicate constraints, *GX* doesn't know which one to use. It can be frustrating but also valuable in terms of thinking about the geometry.

Solution

STEP 1 - The angle is 90°

STEP 2 -

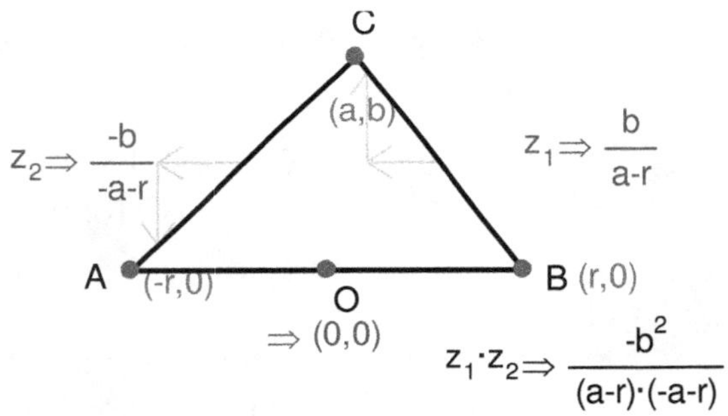

$$OC = \sqrt{a^2+b^2}$$
$$= r$$
$$\therefore r^2 = a^2+b^2$$

$$\text{Slope of AC} = \frac{b}{a+r}$$

$$\text{Slope of BC} = \frac{b}{a-r}$$

$$\text{Product of slopes} = \frac{b}{a+r}\times\frac{b}{a-r}$$
$$= \frac{b^2}{a^2-r^2}$$
$$= \frac{b^2}{-b^2} = -1$$

$\therefore$ AC $\perp$ BC

$\therefore$ The angle inscribed in a semicircle is a right angle.

LAB # 43 *The law of cosines*

Aim: To prove the law of cosines.

This proof of the law of cosines uses coordinate geometry results. There are other methods that could be used.

RTS: $c^2 = a^2 + b^2 - 2ab \cos C$
Where a, b, c are sides of any triangle and C is the angle opposite side c.

Draw the diagram

Draw triangle ABC
Constrain C to (0, 0)
Constrain B to (a, 0) (note this is *a* units long, is opposite vertex A and is often labeled as *a*)
Constrain A to (d, h)

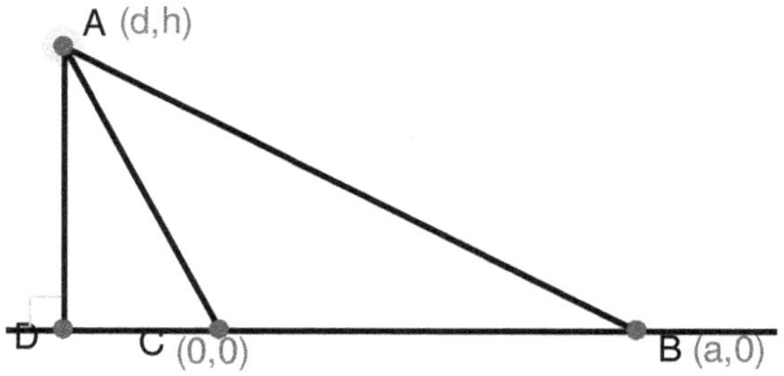

Draw | Infinite Line BC
Construct a perpendicular to BC passing through A.
Mark intersection D of perpendicular with BC.
Hide perpendicular line and Infinite line BC
Draw line segments AD and CD.

Calculate (Output) length
AB, CD, BD and AC.

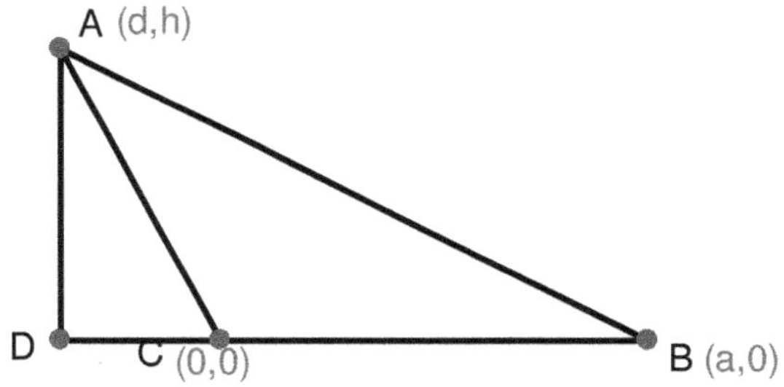

BC = a
AD = ______

AC (b) = ________

AB (c) = ________

Complete the proof:

c^2 =

Draw the diagram

Construct a triangle
Constrain | Length for AC and BC
Constrain | Angle C.
Calculate | Length of the third side (AB).
Draw | Expression to calculate AB^2.

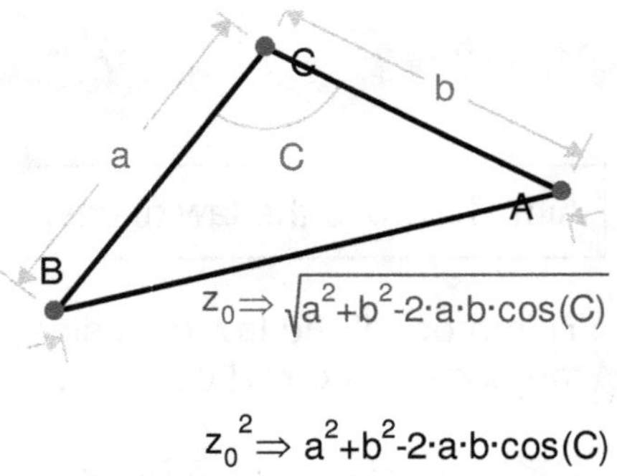

Complete the rearrangement of the law of cosines to make C the subject.

$c^2 = a^2 + b^2 - 2ab\cos C$

$2ab\cos C =$ ______________

$\cos C =$ ________________

C = ____________________

This is used to find the angle in a triangle knowing the three sides.

Reset the *Constraints* on your triangle to sides of a, b and c.

to check your answer for C.

Use your diagram to find expressions for

$\cos A =$ ________________

$\cos B =$ ________________

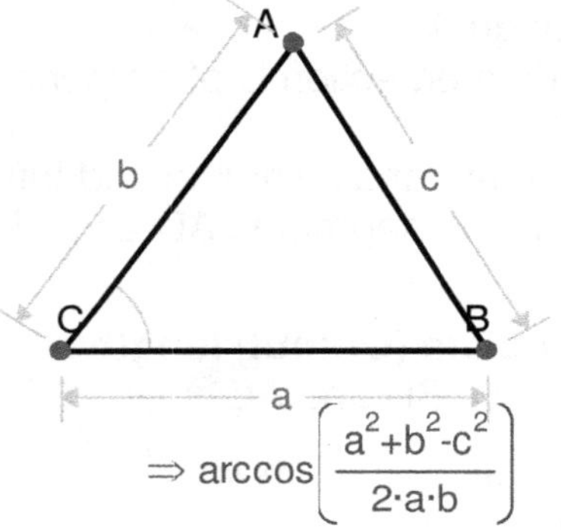

LAB #43 The law of cosines

In this activity students:

- Use GX to measure parts of the construction and
- Complete a proof for the law of Cosines

This lab develops a coordinate geometry proof for the law. Students require skills with algebraic manipulation in order to successfully complete the lab activity. GX enables some of the calculations to be performed. There is not a lot of work required of the students in this lab. It is likely that students will need support to fully understand the last part of the proof, particularly if their algebra skills are not strong.

The final part of the Lab demonstrates how GX uses the law of cosines, producing the formula in symbolic form.

Solution

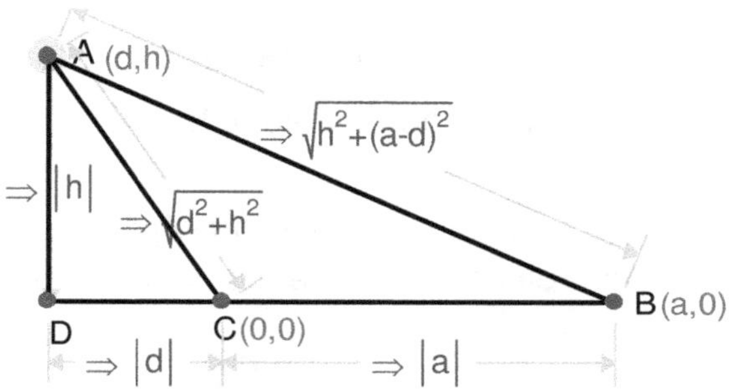

AB (c) = $\sqrt{(a-d)^2+h^2}$

AC (*b*) = $\sqrt{d^2+h^2}$

BC = a

AD = h

$$\begin{aligned} c^2 &= (a-d)^2+h^2 \\ &= a^2-2ad+d^2+h^2 \\ &= a^2-2ad+b^2 \\ &= a^2+b^2-2a(b\cos\mathrm{C}) \end{aligned}$$

$$c^2 = a^2+b^2-2ab\cos C$$

$$2ab\cos C = a^2+b^2-c^2$$

$$\cos C = \frac{a^2+b^2-c^2}{2ab}$$

$$\mathrm{C} = \arccos\left(\frac{a^2+b^2-c^2}{2ab}\right)$$

$$A = \arccos\left(\frac{b^2+c^2-a^2}{2bc}\right)$$

$$B = \arccos\left(\frac{a^2+c^2-b^2}{2ac}\right)$$

LAB # 44 *Subtraction formula*

Aim: To prove the addition formula for cos($A - B$).

Draw diagram

Draw triangle ABC

Constrain A to (0,0)

Draw point D and constrain to (1, 0)

Constrain the angle between the x-axis (AD) and AB to θ

Constrain the angle between the x-axis (AD) and AC to ϕ.

Constrain AB and AC to 1

Calculate(Output)| ∠BAC

Calculate coordinates of B and C

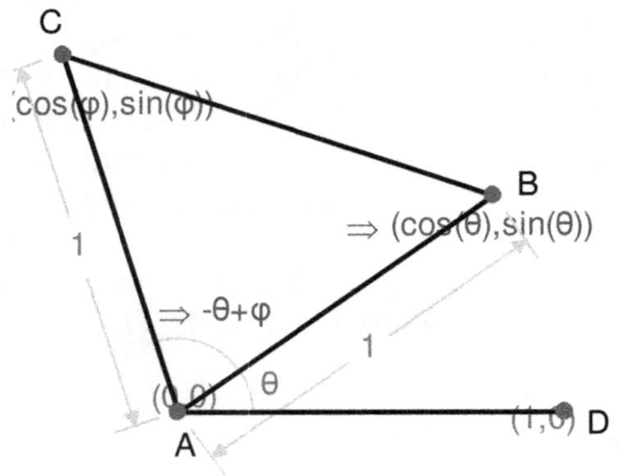

Draw a diagram showing why the coordinates of B are (cosθ, sinθ).

Use GX to Calculate | Length of BC

BC = ______________________________________

What rule/law has GX used to calculate the length? ______________________

Complete the proof:

$BC^2 = (\cos\theta - \cos\phi)^2 + (\sin\theta - \sin\phi)^2$ (Distance formula)

$= \cos^2\theta - 2\cos\theta\cos\phi + \cos^2\phi +$ ______________________

$= \cos^2\theta + \sin^2\theta + \cos^2\phi + \sin^2\phi -$ ______________________

= 1 + 1 – ______________________

= 2 – ______________________

Using the law of cosines $BC^2 = 2 - 2\cos(\phi - \theta)$

$2 - 2\cos(\phi - \theta) = 2 -$ ______________________

$\therefore -2\cos(\phi - \theta) = -2$ ______________________

$2\cos(\phi - \theta) =$ ______________________

$\cos(\phi - \theta) =$ ______________________

EXTENSION

Substitute $-B$ for θ and A for ϕ to derive a formula for $\cos(A + B)$.

Note: $\cos(-B) = \cos B$ and $\sin(-B) = -\sin(B)$

LAB #44 Subtraction formula

In this activity students:

- Use GX to calculate measures in the Cartesian or coordinate plane and
- Use these measures to complete a proof for $\cos(A - B) = \cos A \cos B + \sin A \sin B$.

Points to note in this activity include:

- setting up the diagram to get the difference between two angles;
- an understanding of polar coordinates or the distance angle; representation of points on the coordinate plane;
- the use of the distance formula and law of cosines to calculate the same length and
- algebraic rearrangement.

The lab could be used prior to exposure to the rule. If so I would recommend numerical examples to make it clear that $\cos(A - B) \neq \cos A - \cos B$.

EXTENSION

Calculation of radical expressions for sin, cos or tan for angles that are multiples of 15°. An example is included in the Lab #29.

Solution

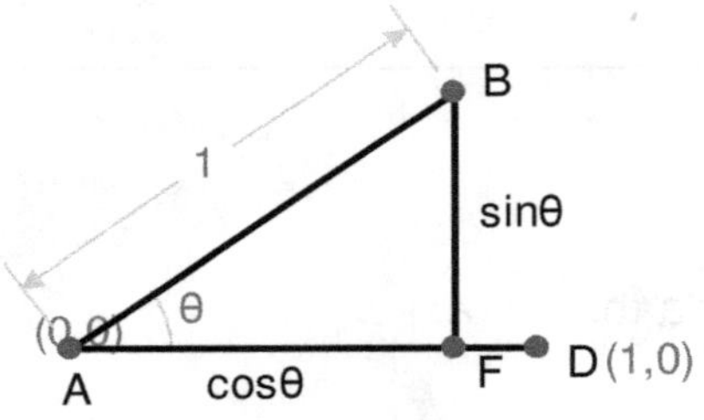

$$\cos\theta = \frac{AF}{1} \Rightarrow AF = \cos\theta$$

$$\sin\theta = \frac{BF}{1} \Rightarrow BF = \sin\theta$$

BC = $\sqrt{2 - 2\cos(\phi - \theta)}$ (Law of cosines)

$$BC^2 = (\cos\theta - \cos\phi)^2 + (\sin\theta - \sin\phi)^2 \quad \text{(Distance formula)}$$

$$= \cos^2\theta - 2\cos\theta\cos\phi + \cos^2\phi + \sin^2\theta - 2\sin\theta\sin\phi + \sin^2\phi$$

$$= \cos^2\theta + \sin^2\theta + \cos^2\phi + \sin^2\phi - 2\cos\theta\cos\phi - 2\sin\theta\sin\phi$$

$$= 1 + 1 - 2\cos\theta\cos\phi - 2\sin\theta\sin\phi$$

$$= 2 - 2\cos\theta\cos\phi - 2\sin\theta\sin\phi$$

Using the law of cosines $BC^2 = 2 - 2\cos(\phi - \theta)$

$$2 - 2\cos(\phi - \theta) = 2 - 2\cos\theta\cos\phi - 2\sin\theta\sin\phi$$

$$\therefore -2\cos(\phi - \theta) = -2\cos\theta\cos\phi - 2\sin\theta\sin\phi$$

$$\cos(\phi - \theta) = \cos\theta\cos\phi + \sin\theta\sin\phi$$

Chapter 10 – Loci

Teacher Notes

Locus: any system of points, lines, etc. which satisfies one or more given conditions.

Constraints support an intuitive way of creating loci. For example, conic sections arise from a rule such as the distance from a fixed point is in a fixed ratio to the distance from a fixed line. In GX this is constructed by constraining a point (the fixed point), constraining a line and then constraining a point to meet the locus requirements. In this case constrain the distance from fixed point to a and the distance from the line to $t * a$.

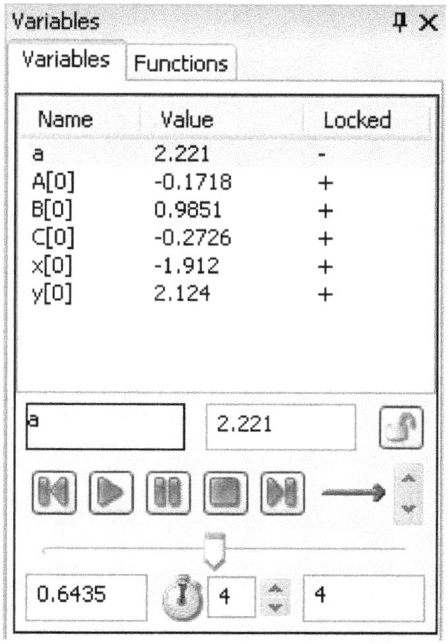

The variable window is potentially very valuable in developing algebraic understanding. An example is displayed on the left.

The window provides a link between numbers and their visual representation in the diagram. It is dynamic and the slider is a great way to assist students to view a variable as something that changes. Animation further enhances this as well as being engaging.

The labs in this chapter involve students in creating the basic drawing and then dragging the point to see how and where it moves. This is intended to support conceptual understanding of loci as a set of points.

Once the path has been explored the locus tool is used to create a trace of those positions. The software will often draw only a portion of the locus, compared to the results of dragging. A class discussion about why this occurs would be beneficial. Another point of discussion is the range allowed for the locus parameter.

This will need to be changed (Edit | Edit Curve Domain, or just double-click the curve) depending upon the curve being considered.

Loci considered in this chapter involve:

- A fixed distance from a given point, the circle;
- Locating the epicenter of an earthquake, intersection of circles;
- Equidistant from two points, perpendicular bisector of a line segment;
- An application using the perpendicular bisector;
- A compound locus problem;
- Curves of conic sections based upon the focus directrix properties and
- Bezier curves or spline curves.

Symbolics can extend the labs by looking at the equations of the loci. Where this has been incorporated into the labs the defining points and lines are constrained to values or equations that produce the standard equations.

LAB # 45 *The circle*

Aim: To introduce the circle as the locus of points equidistant from a fixed point

Find all points that are 2 units away from a given point.

Plot two points
  Draw | Point
  Click in two places

Set the position of A to the origin and AB to 2 units
  Select Tool
  Select A Constrain | Coordinate
  Enter (0,0)
  Select points A and B
  Constrain | Distance/Length
  Enter 2

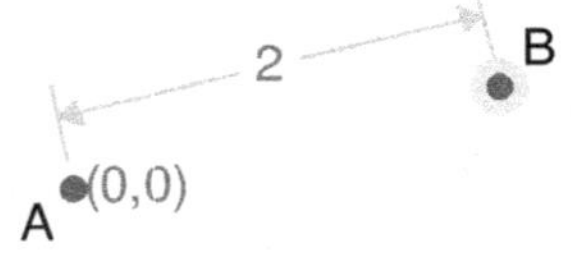

Drag B around.

What path does B follow? ________________________________

Animate B

Draw the line segment AB
  Select the line segment tool
  Click on point A
  Click on point B

Constrain the angle of line AB
  Select tool
  Click on the line
  Constrain Direction
  Press enter (accept the default value θ)

Play animation using θ, the driving variable
  Variables toolbox
  Select *θ*
  Press the play button
  Experiment with changing values in the bottom part of the window

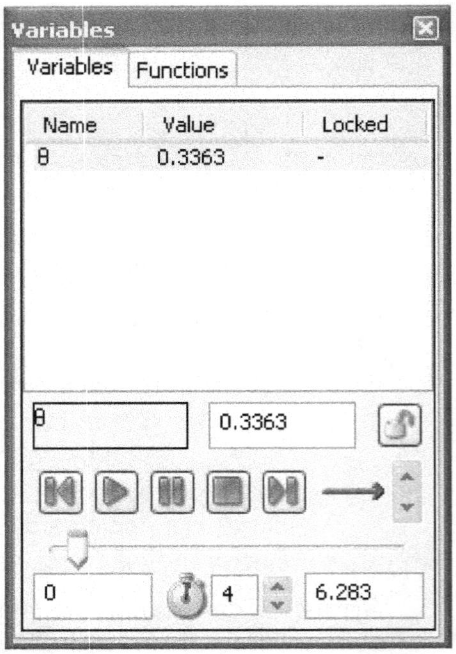

Sketch the locus.

Select B

Construct | Locus .

Set the parametric value to θ, the direction of the line.

Set the start value to 0.

Set the end value to 360 or (6.28 if in radian mode).

Click OK

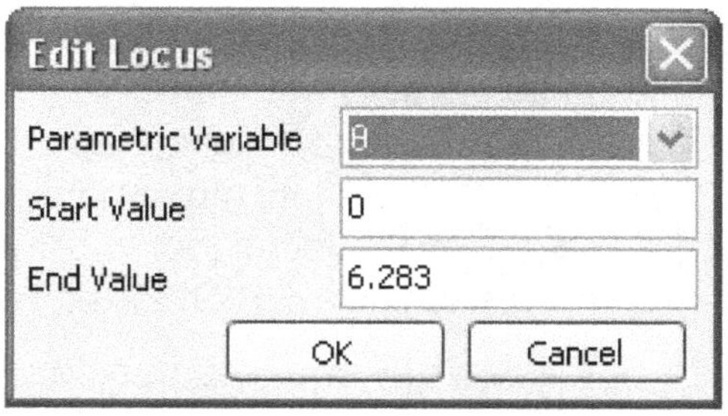

(To set the angle to degrees: Edit | Settings | Math | Angle Mode | Degrees)

Describe the locus ______________________________

Complete the definition; A circle is the set of points ______________________________

What is the minimum information that is needed for two people to construct identical circles in the same position on the coordinate plane?

____

EXTENSION: Equation of the circle

Use GX to find equations for the following circles.
Select the circle
Calculate | Explicit equation.

a. The equation of the circle center (0, 0) and radius 2 is ______________________________

b. Change the radius to 5
Double click on the distance AB and change 2 to 5
The equation of the circle center (0, 0) and radius 5 is ______________________________

c. Change the radius to *r*
The equation of the circle center (0, 0) and radius *r* is ______________________________

d. Double click on (0, 0), the coordinates of A. Change A to (0, 1).
The equation of the circle center (0,1) and radius 2 is ______________________________
Verify that it is the same as $x^2 + (y-1)^2 = 4$

e. Change A to (-2, 1)
The equation of the circle center (-2, 1) and radius 2 is ______________________________
Verify that it is the same as $(x+2)^2 + (y-1)^2 = 4$

f. The equation of the circle center (-2, 1) and radius 3 is ______________________________
Verify that it is the same as $(x+2)^2 + (y-1)^2 = 9$

g. The circle $(x-h)^2 + (y-k)^2 = r^2$ has center __________ and radius __________

LAB #45 The circle

In this activity students

- Drag a point, constrained to a fixed distance from the origin, to trace a circle.
- Animate this movement
- Use the locus tool to draw the circle.
- The EXTENSION explores the equation of the circle.

This activity could be used as an introduction to and familiarization to GX. Setting up the parameters of the locus, i.e. a fixed distance from the origin is straight forward to do using Constraints. The structure of the Lab asks students to drag and visualize, then animate before using the locus tool. Such an approach can enhance student understanding.

The EXTENSION explores the general form of the equation of a circle using *Symbolics*. This could be done as a separate Lab at another time when looking at the equation of the circle. GX's capacity to generate such expressions is increasingly valuable as students develop algebraic thinking and proficiency. Algebraic manipulation skills are required to convert GX's expression (general conic form) to the center, radius form.

Solution

B moves in a circle.

A circle is the set of points equidistant from a point.

The center of the circle and the radius are required to specify a circle.

EXTENSION

center	radius	equation
(0, 0)	2	$-4+x^2+y^2=0$ or $x^2+y^2=4$
(0, 0)	5	$-25+x^2+y^2=0$ or $x^2+y^2=25$
(0, 0)	r	$x^2+y^2-r^2=0$ or $x^2+y^2=r^2$
(0,1)	2	$-3+x^2-2y+y^2=0$ or $x^2+(y-1)^2=2$
(-2, 1)	2	$1+4x+x^2-2y+y^2=0$ or $(x+2)^2+(y-1)^2=4$
(-2, 1)	3	$-4+4x+x^2-2y+y^2=0$ or $(x+2)^2+(y-1)^2=9$

The circle $(x-h)^2+(y-k)^2=r^2$ has center (h, k) and radius r.

LAB # 46 *Shake it Mama*

Aim: To locate the epicenter of the earthquake.

Earthquake Data from the US Geological survey Wednesday, Nov 30, 2005

Use the grid on the map (next page) to complete the table.

Seismic (earthquake recoding) station	Coordinates	**Distance from epicenter**	
		Km	(grid units)
Hilo, Hawaii :	(,)	45	0.45
Mauna Loa, Hawaii :	(5.5 , 2.5)	100	
Lahaina, Maui :	(5, 4)	165	
HONOLULU, Oahu :	(,)	400	

Open a new drawing in GX

Plot the recording stations
  Draw | Point
  Constrain | Coordinates

Repeat for each recording station

Draw a circle center Hilo of radius 0.45 units. (The epicenter was 45 km from Hilo.)
  Circle tool
  Click on Hilo
  Move the mouse and click again (or drag and release)
  Select the circle
  Constrain | Radius 0.45

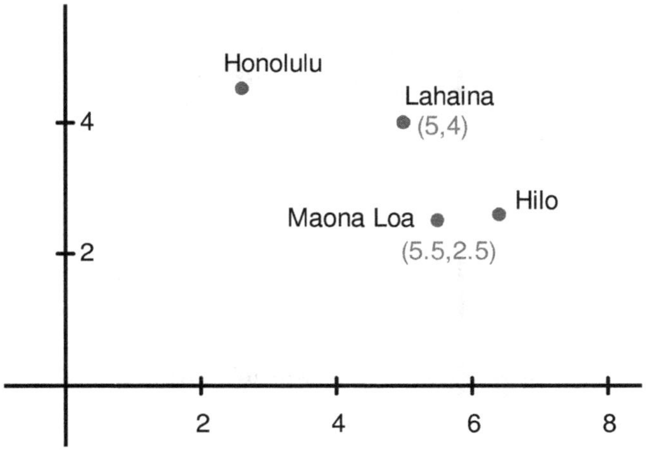

Repeat for Mauna Loa and Lahaina

The coordinates of the earthquake's epicenter are (______ , ______.)

Add Honolulu to your diagram and draw a circle of radius 4 units.

Does this support the location determined from the other three stations?______________

____

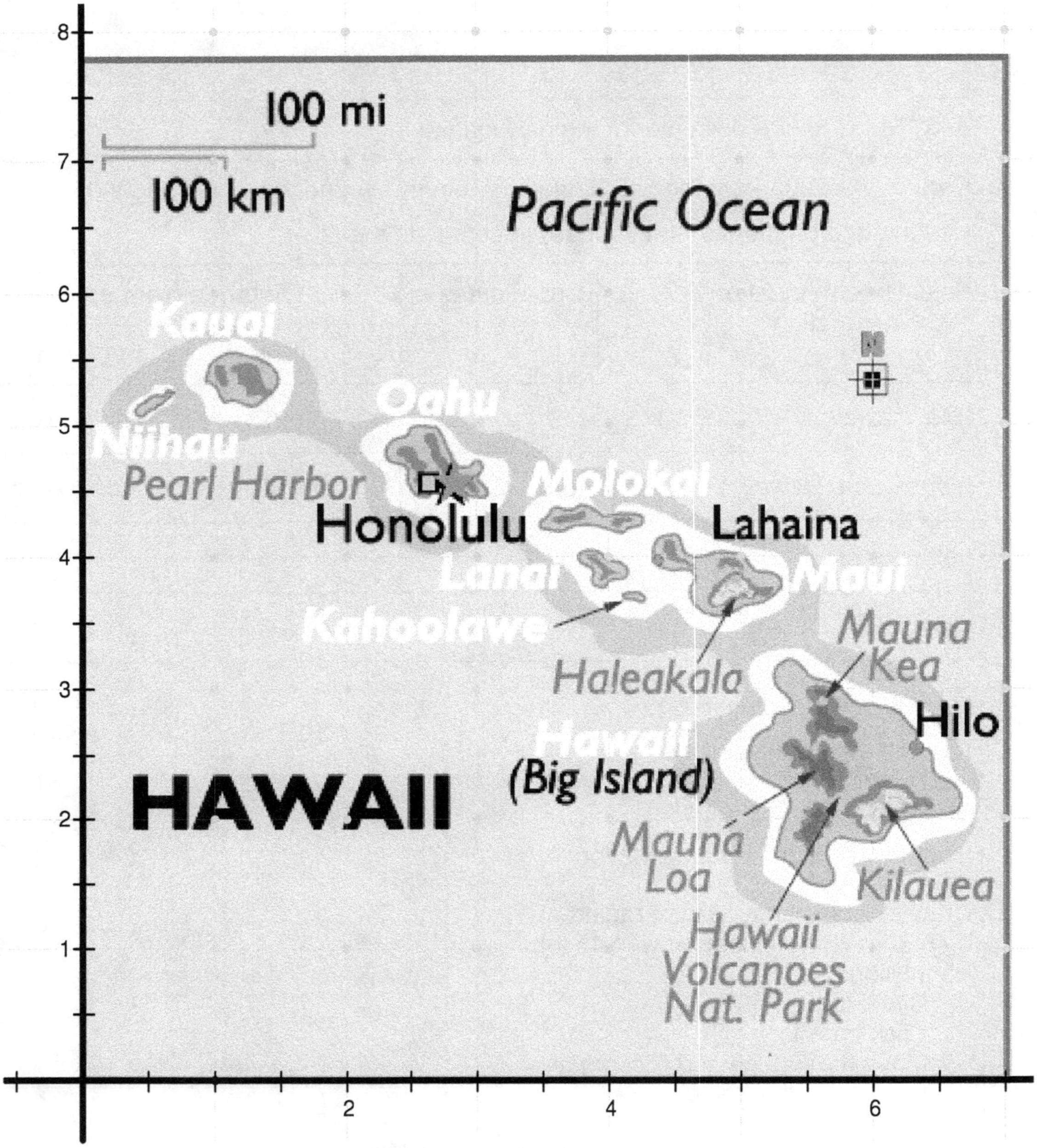

Scale: 1 grid unit = 100 km

LAB #46 Shake it Mama

In this activity students:

- Locate the earthquake recording stations on a grid;
- Plot these on the coordinate plane and
- Find the intersection of the circles to locate the epicenter.

Real data from an Hawaiian earthquake provides the context for this activity. The earthquake data gives the distance of the earthquake's epicenter from a number of seismic stations in the region. The activity may be introduced with a discussion of earthquakes, particularly considering any recent major seismic events. It may be linked to science and social studies.

A circle of this radius restricts the possible location and the intersection of the circles locates the epicenter. *Geometry Expressions* is used by placing a grid over the map and using the resulting coordinates to constrain the recording stations. The activity could also be very effectively done on paper using drawing instruments.

The epicenter is the point on the Earth's surface directly above the point where the earthquake originates. In practice the locus rarely locates a point, rather a small region. Another source of error to be considered is the accuracy with which the grid coordinates of the seismic stations are located. In practice these would be known very accurately.

Solution

Seismic station	Coordinates	Km	(grid units)
Hilo :	(6.4, 2.6)	45	0.45
Maona Loa, Hawaii :	(5.5 , 2.5)	100	1
Lahaina, Maui :	(5, 4)	165	1.65
HONOLULU, Hawaii :	(2.6, 4.5)	400	4

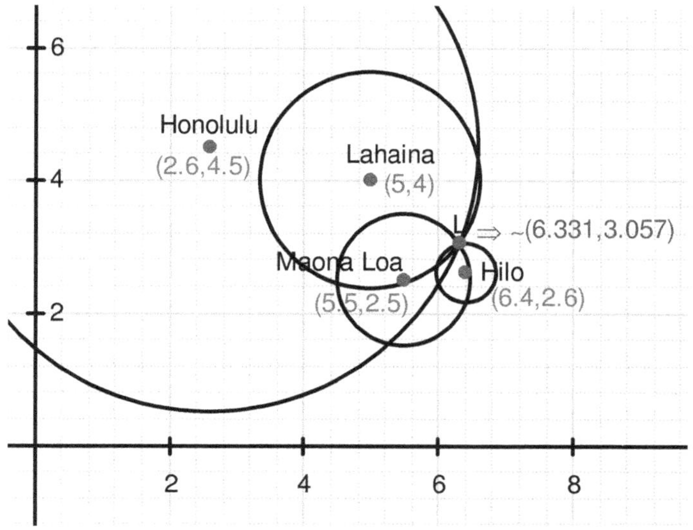

Epicenter at (6.3, 3)

LAB # 47 *Distance from two points*

Aim: To explore the locus of points equidistant from two points.

Part 1 – Create the locus.

Plot 3 points

Fix the points A and B
- Select the point
- Constrain | Coordinates
- accept the default values

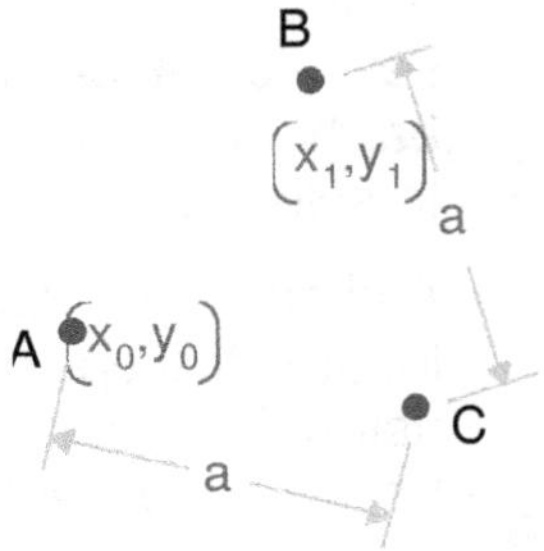

Constrain AC and BC to the same distance apart
- Constrain | Distance
- Enter *a*

Drag C around

Describe the possible position of C as you drag the point around ____________________

____

Look at the variables window. While dragging C around look at the range of values that *a* can take. Make a statement about the minimum possible value for *a*.

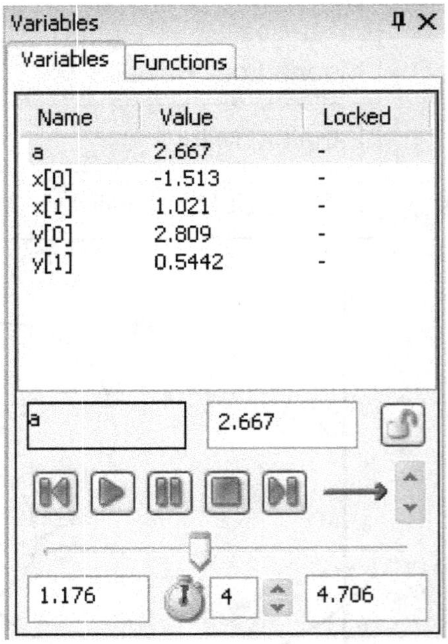

____

____

____

Create the locus
- Select C
- Construct | Locus and accept the default values

Part 2 - Describe the locus

Draw the line segment AB
Place a second point D on the locus
Hide the locus
Draw an infinite line through CD

Mark the intersection E of the lines
Select lines AB and CD
Construct | Intersection

Drag C around, drag A and B around

Measure the angle between AB and CD ______________________

What do you notice? ______________________

____

EXTENSION

Delete points D and E. Delete line CD

Construct a reflection of the locus using AB as the mirror line.
View | Show All – to unhide the locus
Select the locus
Construct | Reflection
Click on line AB

Change the length of AC to 2 * a. What is the locus now?

____

Experiment with other multipliers. Describe your findings.

____

____

____

____

Generalize with t * a.

____

____

LAB #47 Distance from two points

In this activity students

- Construct the locus
- Explore the properties and describe the locus

The locus is the perpendicular bisector of the line segment joining the two points. The construction of the line and intersection is necessary to be able to measure the distances and angle. The distances should be Real (decimal) measurements. An important part of the lab is for students to play with the figure and drag it around to deepen understanding of the geometry.

There is scope for a rich discussion about the possible values for *a.* many students will just imagine that a is a real number and are most surprised that there is a minimum equal to half the distance between the two fixed points. For example questions like; can you make *a* negative to get the rest of the line (no!) Why? How do you get the rest of the line?

The Extension shows how small changes in the parameters can make significant changes in the resulting locus. A discussion about the need for a reflection may be required. It would be nice if GX understood what we wanted here. The description is open ended and it would be possible to ask the better students to describe in more detail including giving expressions for the radius and center of the circle.

Another Extension for the first part of the Lab would be to look at the equation of the line. GX can just give the equation, however students will learn more if they build towards it by constraining the end points to simple values such as two points on an axis to begin with.

Solution

C lies on a line, the perpendicular bisector of AB.

Answers will vary. The possible values for a are $a \geq$ AB / 2

AB $\perp$ CD for any positions of A and B.

EXTENSION

For the length of AC set to $2*a$ the locus is a circle

As the value of the multiplier changes, the size of the circle changes.

LAB # 48 *Detective Work?*

Aim: Use locus to investigate a crime scene.

Detective Gray is investigating a crime in Central Park and he suspects the Brown brothers, Bugs, Tom and Gerry, of being the perps.

Witness reports

Person	Time	Location	Suspect	Action
A lady	2:10pm	Warrior's gate ON Central Park North	Gerry	Running north
Street vendor	2:10pm	Central Park West and 106th Street	Bugs	Running into 106th St.
Officer Ginger	2:10pm	Bridge on West Drive	Tom	Running South

Detective Gray thinks they were all running away from the crime.

"If they were together at the crime and then separated, we can determine the meeting place. We assume that they have moved the same distance from the crime scene and have run in a straight line."

He asks you, as his math expert, to determine their point of separation. (Where they were when they were together.) If this matches the crime scene, then he will arrest them.

Draw a grid over the map (next page) so you can plot the positions of the witnesses using GX.

1. Determine the coordinates of the witnesses.
2. Plot these positions in a GX drawing.
3. Determine where Tom and Gerry could have been together and mark this on the map. (Tom's distance from Officer Ginger is the same as Gerry's distance from the lady at Warrior's gate.
4. Determine where Tom and Bugs could have been together.
5. Determine where Bugs and Gerry could have been together.
6. What is your answer for Detective Gray?
   Of course this needs to be backed up with reasoning that will stand up in court!

____

____

____

Mark your answer on the map below.

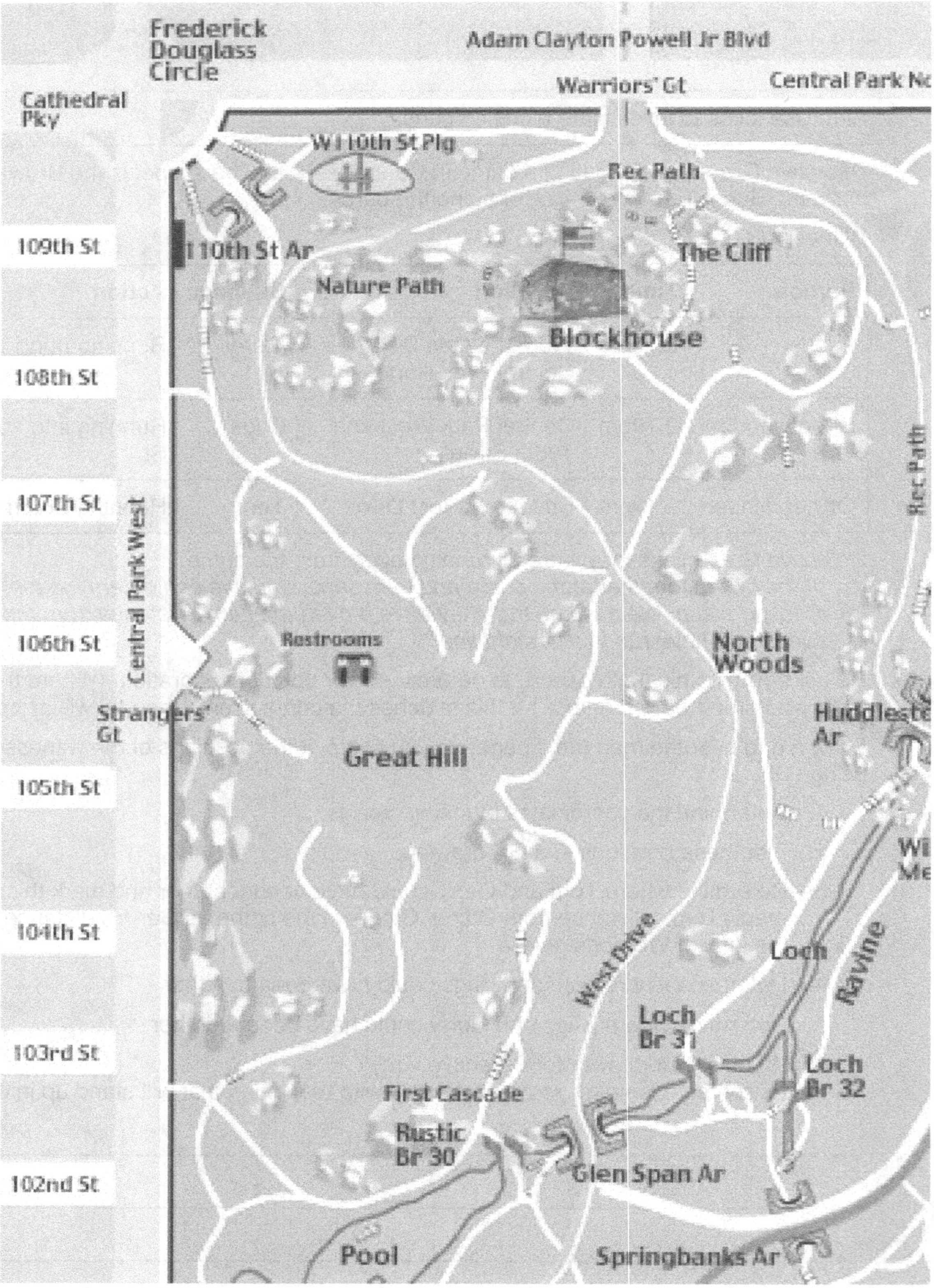
Frederick
Douglass
Circle
Adam Clayton Powell Jr Blvd
Cathedral
Pky
Warriors' Gt
W 110th St Plg
Rec Path
109th St
110th St Ar
Nature Path
The Cliff
Blockhouse
108th St
Rec Path
107th St
Central Park West
106th St
Restrooms
North
Woods
Strangers'
Gt
Great Hill
105th St
104th St
West Drive
Loch
Ravine
Loch
Br 31
103rd St
Loch
Br 32
First Cascade
Rustic
Br 30
102nd St
Glen Span Ar
Pool
Springbanks Ar

LAB #48 Detective Work?

In this activity students:

- Interpret the scenario;
- Apply a coordinate system to the picture and
- Solve the problem using the perpendicular bisectors

Building on from the previous activity students can use the perpendicular bisector to help Detective Grey. The context, while contrived, is set in New York's Central Park and provides an opportunity to apply locus concepts in a non-routine situation.

From witness statements Detective Gray thinks the three brothers were all running away from the crime. "If they were together at the crime and then separated, we can determine the meeting place. We need to assume that they have moved the same distance from the crime scene." That is the two brothers could be anywhere on the perpendicular bisector of the line between their sightings.

For the adventurous it would be possible to Draw | Picture and import the image file. The challenge will be to maintain the aspect ratio as GX fits the image file into the rectangular area chosen. It is a good question to explore what effect this has on the answer.

The locus is the circum center of the triangle formed by the positions of the three observers. This is a connection you might like to explore further.

The activity can be done as a pencil and paper exercise using drawing instruments.

Solution

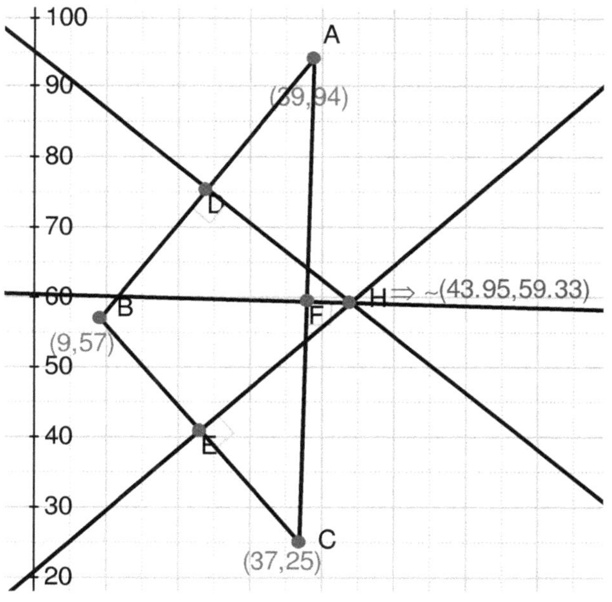

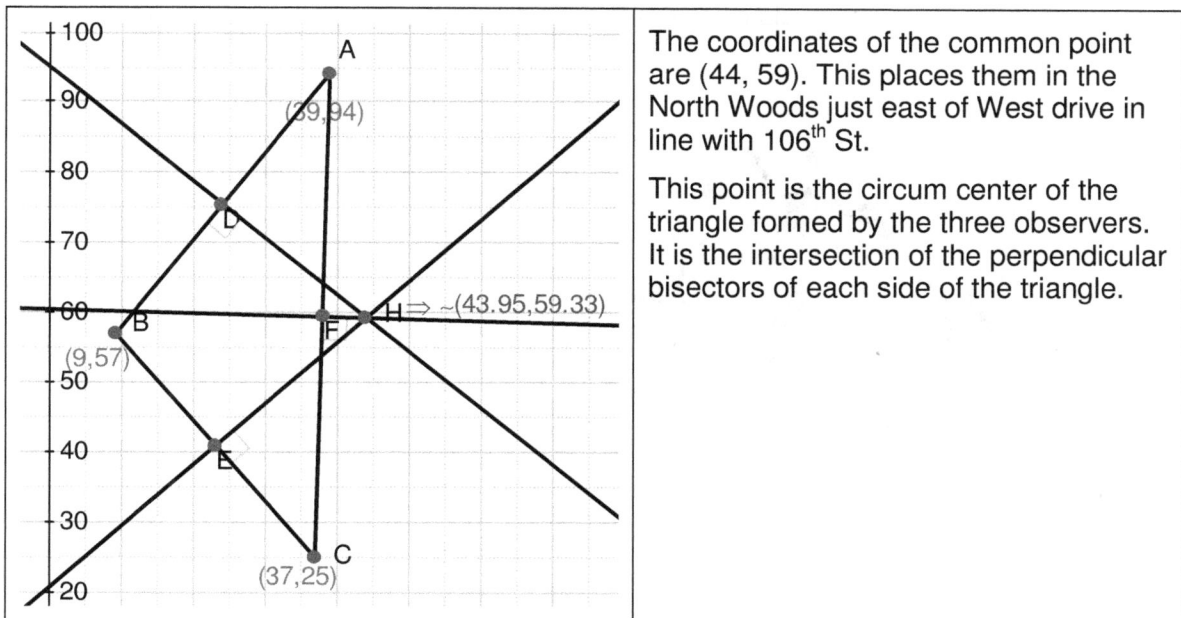

The coordinates of the common point are (44, 59). This places them in the North Woods just east of West drive in line with 106th St.

This point is the circum center of the triangle formed by the three observers. It is the intersection of the perpendicular bisectors of each side of the triangle.

LAB # 49 *Turkey tether*

Aim: To solve a compound locus problem.

Have you ever decided to fatten your own turkey for Thanksgiving? Me neither, but I thought I'd write a problem as if I had.

There is a limited amount of room in my yard, given the gardens, the fence, and the neighbors' boat (don't ask). I decided my turkey could live between a tree, a covered food area, and a little pond. The tree is 8 feet from the pond and 9 feet from the food. I've heard that you can't put the food and water too close together or the turkey will drop the food into the water, so the food and pond are 10 feet apart. I've got a perfect piece of rope that will give my bird a 6 foot range from wherever I anchor the rope.

1. *Where should the anchor be so that it's an equal distance from the tree, the water, and the food?*
2. *Will the six foot range be enough for the turkey to reach everything he needs (food, water, and shade)?*

(Fetter 1995)

EXTENSION

Describe how you would do the problem using drawing instruments pencil and paper.

Math Forum
http://mathforum.org/pow/solutio44.html

LAB #49 Turkey tether

In this activity students:

- Create their own *Constraints* to represent the problem on the coordinate plane;
- Draw the diagram using GX and
- Interpret the solution.

This is a compound locus problem. The setting is a turkey tethered in a yard and determining the area that can be grazed. The problem is derived from one presented on the Math Forum.

The lab does not provide instructions for creating the drawing so students are required to use previously developed skills. The solutions show three different methods. To extend students questions such as *In what region can the tether be fixed?* and *What is the minimum length of rope?* Can be used to extend students thinking.

Solution

Method 1

Place a point inside the triangle, measure distances to each amenity and drag the point to meet the conditions.

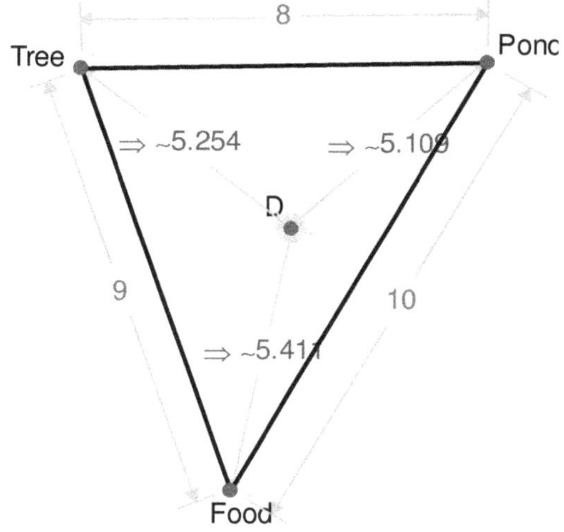

Method 2

Draw circles of radius 6. The stake can be placed in the region where the three circles overlap.

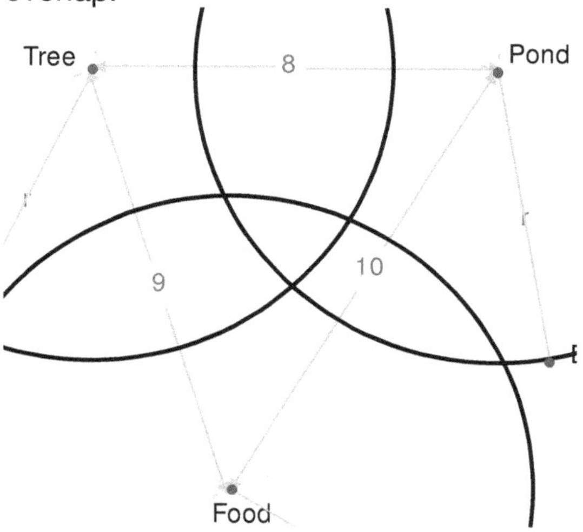

Method 3

Draw the perpendicular bisectors of each side of the triangle. This point is the circum-center of the triangle. The radius of the circum circle is the minimum length of rope required.

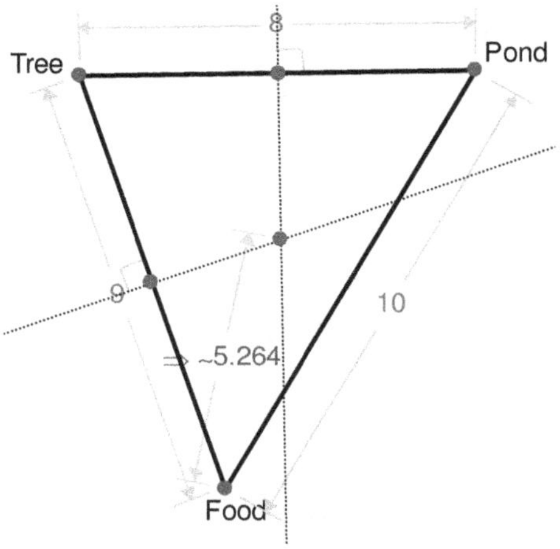

LAB # 50 *Parabolic focus*

Aim: To explore loci of conic sections based on focus and directrix.

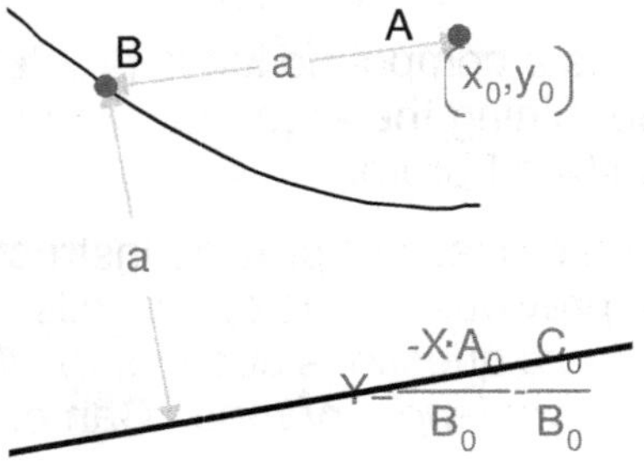

Create the drawing in GX.

Draw two points and a line

Constrain | Distance/Length of point B from the line to the distance *a*

Constrain | Distance/Length of Point A from point B to the same value, a.

Drag B around to explore the locus of points equidistant from a line and a point.

Constrain the Coordinates of point A. Use the default values.

Constrain the Implicit equation of the line. Use the default values.

Make a sketch in the box tracing possible positions for B.

Create the locus of B

Select B

Construct | Locus

From the dialog box;

Choose *a* for the Parametric Variable.

Set Start Value (e.g. 0)

Set End Value (e.g. 5)

These values can be changed later by right clicking on the locus and choosing Edit Curve Domain.

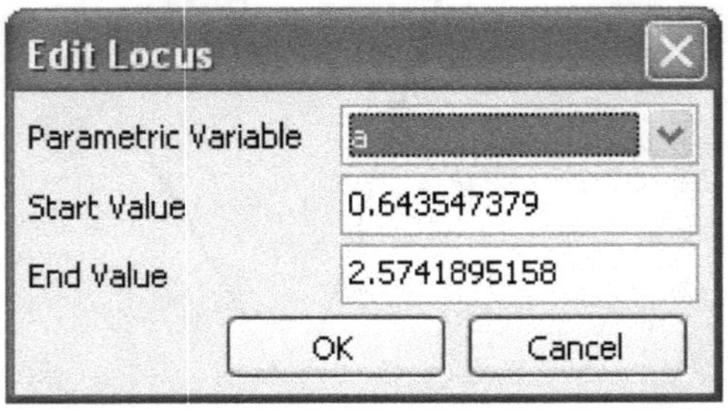

Note: Only one half of the locus is formed. To see the other half reflect the locus in the perpendicular through A.

Select A and the line

Construct | Perpendicular

Select locus

Construct | Reflection

Click on the perpendicular line

Animate B

To animate B the value of *a* must vary. The Variables toolbox contains the animation controls.

Select *a* in the Variables toolbox

Press the play button and experiment with the buttons and controls.

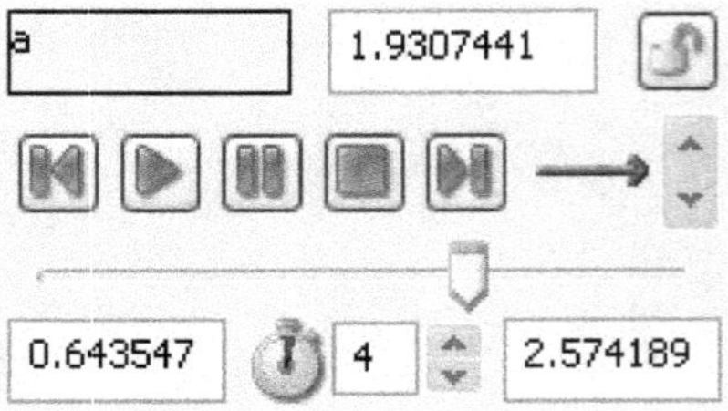

Constrain AB to $t*a$

Constrain point A to (0, 1) and line to $y + 1 = 0$

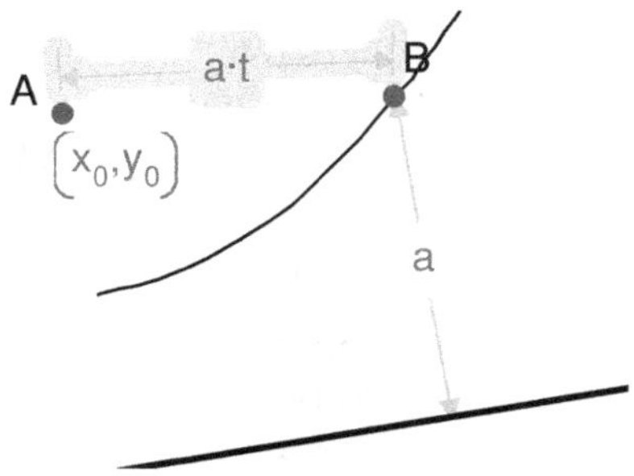

Note two possible forms of the equation: Edit | Settings | Math | Calculation | Line Equation Style – use the first one:

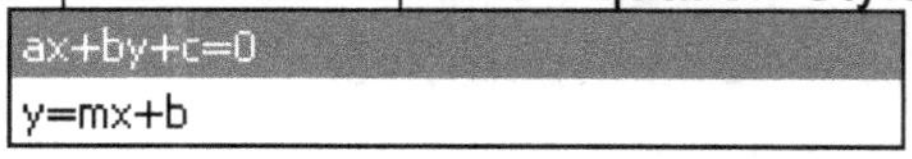

Explore the locus for different values of t.

Select t in the Variables toolbox and move the slider.

Record sketches of some different shaped loci.
To change the domain of the locus: Double-click the locus or select it and choose Edit | Curve Domain.

EXTENSION

Reset the distance AB to a.

Determine the equation of the locus: 

Select the locus

Calculate | Implicit Equation

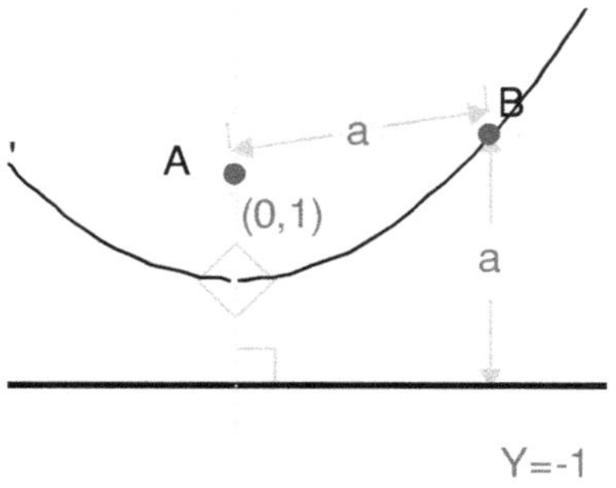

____

Determine the equation for focus at (0, *b*) and directrix (line) $y = -b$

____

Determine the equation for focus at (b, 0) and directrix (line) $x = -b$

____

Explore the equation of the locus for the focus at

(*b*e*, 0) and directrix $y = \frac{-b}{e}$.

Describe your discoveries. ____

____

____

____

LAB #50 Parabolic focus

In this activity students:

- Construct a drawing with focus and directrix;
- Construct the locus, a parabola;
- Animate the locus and
- Alter the *Constraints* to produce ellipses and hyperbolas.

This lab explores conic sections arising from the focus – directrix property. Students should be comfortable using the software to be able to focus on the mathematics in this lab. The ease with which it is possible to change the locus by altering the ratio of distance from focus to distance from directrix is stunning. Encourage students to use the variable window to change the values of variables. Dynamically this can be done using the slider.

The Extension explores equations of the loci. The capacity of GX to constrain the point to particular coordinates and the equation of the line makes the transition to a symbolic representation easy to explore. This would be excellent support to work where students were generating such equations algebraically.

Solution

The parameter t is the eccentricity of the conic section.

<table>
<tr><th>t < 1 - ellipse</th><th>t = 1 - parabola</th><th>t > 1 - hyperbola</th></tr>
<tr><td>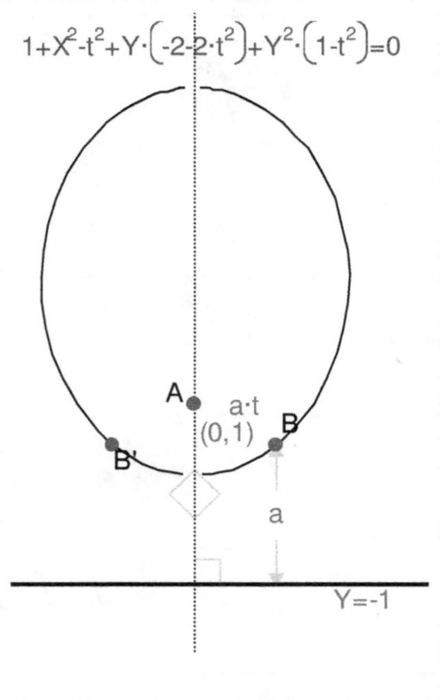
</td><td>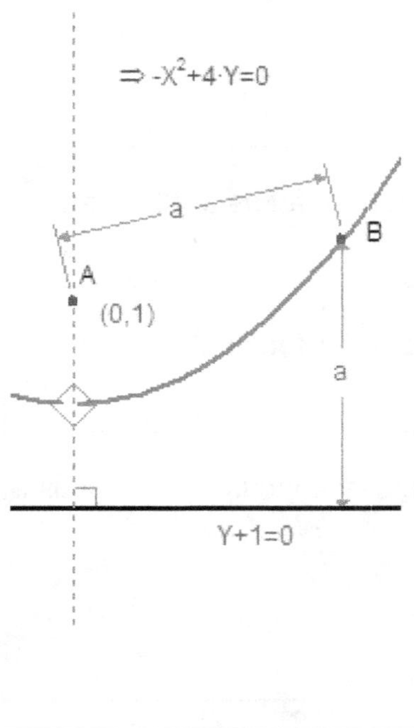
</td><td>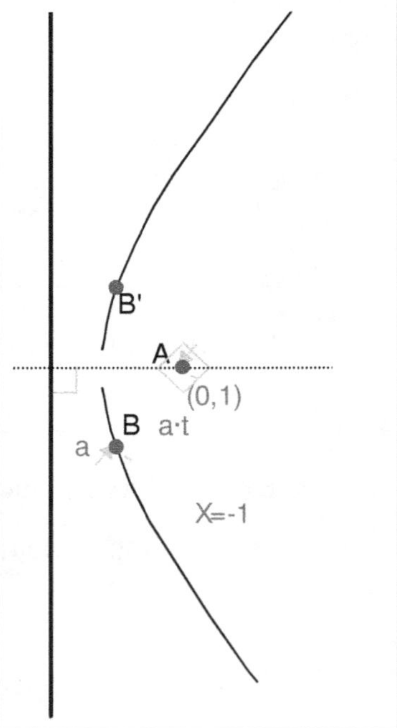
</td></tr>
</table>

EXTENSION

$-x^2 + 4y = 0$, $-x^2 + 4by = 0$, $-y^2 + 4bx = 0$

$$x = be + \sqrt{b\left(\frac{b}{e^2} + \frac{2y}{e}\right)}$$

LAB # 51 *Bezier curves*

Aim: To create and explore Bezier or cubic spline curves as loci.

Part 1 Draw a bezier curve

Open a new window

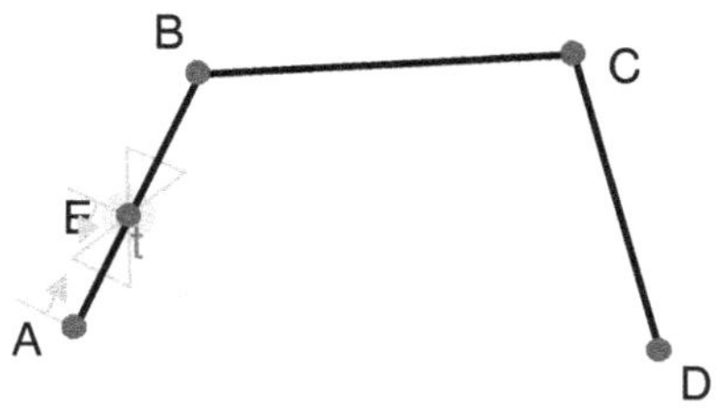

Construct three line segments Place a point E on segment AB.

Select the point E and line segment AB (ctrl click)

 Constrain | Point proportional along curve with the value *t*.

(Note: As E moves along AB, *t* takes values from 0 to 1)

Repeat for BC and CD creating points F and G with the same value, t, so that as E is dragged along AB, F and G will move too.

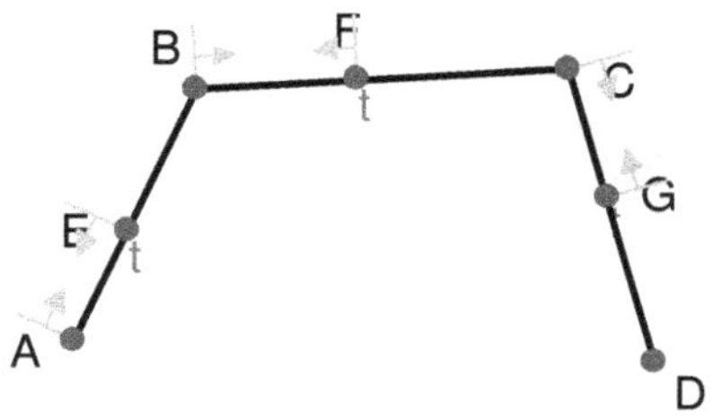

Construct lines EF and FG.

Place points H and I on the segments and set the point proportional constraint to *t*.

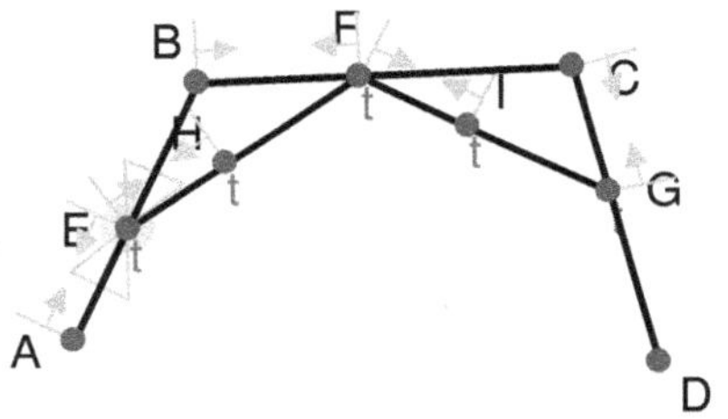

Construct HI

Place point J on HI and set the point proportional constraint to *t*

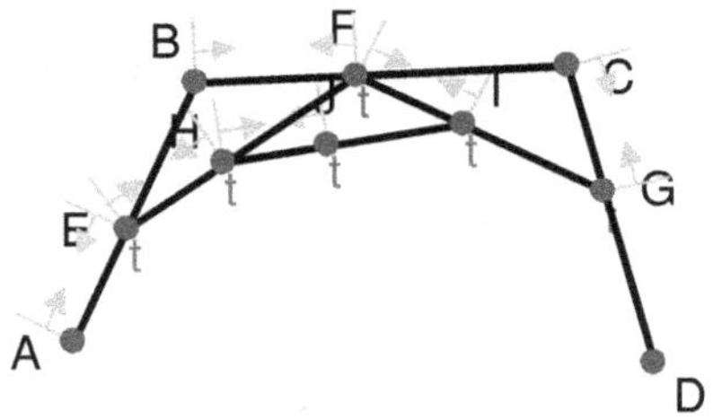

Construct the locus of J 

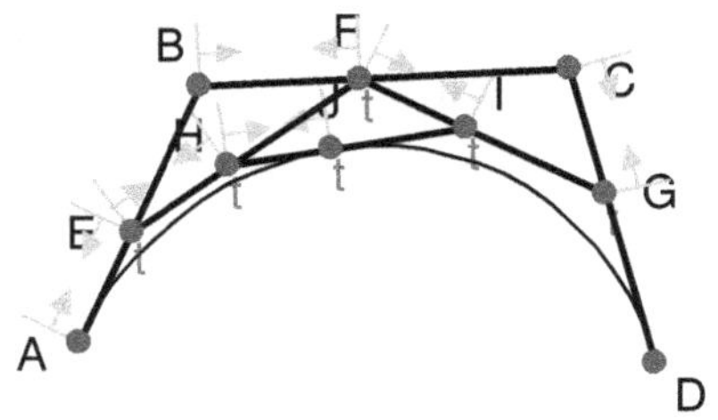

Hide the construction lines and all points *except* B, C, A, D, and J.

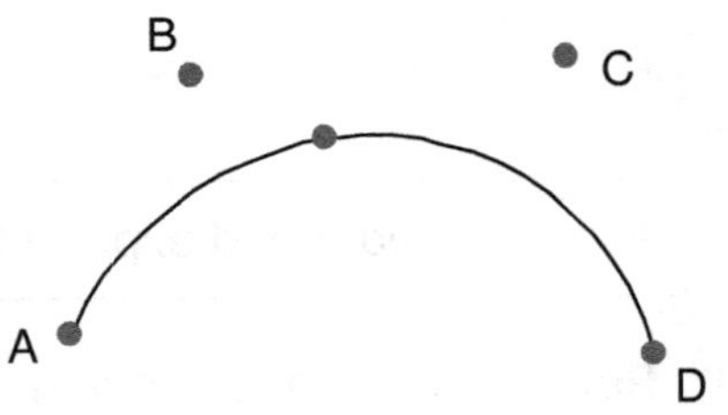

Experiment with dragging the control points A, B, C and D to explore the curves that can be created.

Save your sketch.
Explore animating point J.

Change the line colour and thickness of the locus to make it stand out.

Select the locus curve,
Right click
Choose properties.

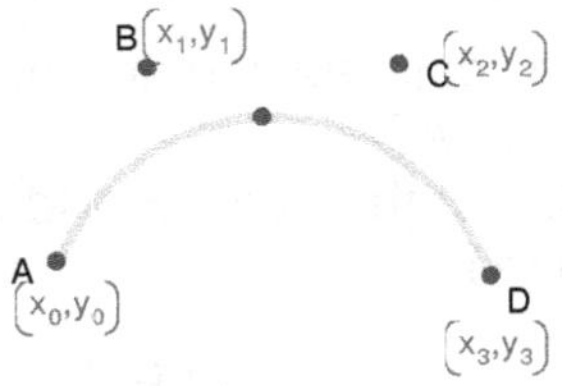

Constrain the coordinates of A to (x_0,y_0)
Constrain B to (x_1,y_1)
Constrain C and D to default values

Select the locus and calculate parametric equation

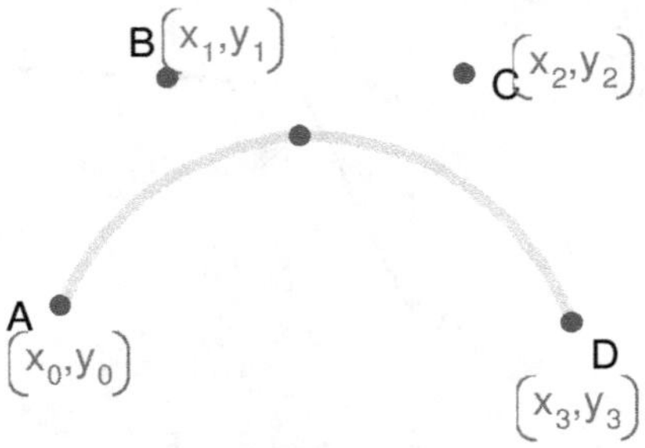

$$\begin{bmatrix} X=x_0-3\cdot t\cdot x_0+3\cdot t^2\cdot x_0-t^3\cdot x_0+3\cdot t\cdot x_1-6\cdot t^2\cdot x_1+3\cdot t^3\cdot x_1+3\cdot t^2\cdot x_2-3\cdot t^3\cdot x_2+t^3\cdot x_3 \\ Y=y_0-3\cdot t\cdot y_0+3\cdot t^2\cdot y_0-t^3\cdot y_0+3\cdot t\cdot y_1-6\cdot t^2\cdot y_1+3\cdot t^3\cdot y_1+3\cdot t^2\cdot y_2-3\cdot t^3\cdot y_2+t^3\cdot y_3 \end{bmatrix}$$

Part 2 Model a Towel Rail

This curve is to be cut out by a Computer Assisted Machining tool (CAM).

A Bezier equation of the curve is required to do this.

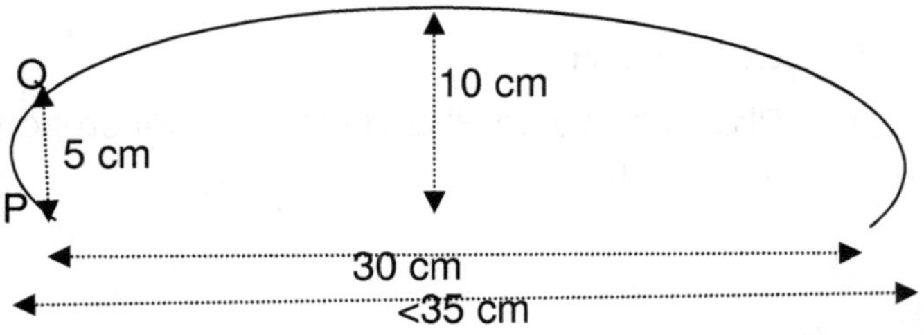

Draw this using your sketch from Part 1.

Hint:

Turn the grid and axes on, click  twice, or select View | Axes and View | Grid
Constrain the end points, A and D, to a gap of 30 units e.g. (0, 0) and (30, 0)
Drag the control points, B and C, until your curve matches the diagram.

Find the coordinates of the control points
Calculate | Real | Coordinates

a. Determine the equation of this curve in parametric form.

> *Note*:
>
> The point $P(x(t), y(t))$ lies on the Bézier curve with initial point (x_0, y_0), intermediate control points (x_1, y_1) and (x_2, y_2) and endpoint (x_3, y_3) if
>
> $$x(t) = a_x t^3 + b_x t^2 + c_x t + d_x$$
> $$y(t) = a_y t^3 + b_y t^2 + c_y t + d_y$$
>
> where $\begin{cases} a_x = x_3 - 3x_2 + 3x_1 - x_0 \\ b_x = 3x_2 - 6x_1 + 3x_0 \\ c_x = 3x_1 - 3x_0 \\ d_x = x_0 \end{cases}$ and $\begin{cases} a_y = y_3 - 3y_2 + 3y_1 - y_0 \\ b_y = 3y_2 - 6y_1 + 3y_0 \\ c_y = 3y_1 - 3y_0 \\ d_y = y_0 \end{cases}$
>
> You can have *GX* display these equations by constraining end points back to the symbolic default values (Part 1 above) and Calculate | Symbolic | Coordinates for point J.

Use these equaions to:

b. Find the maximum width.
c. The slope of the curve at points P and Q.
d. The equation of the tangent through P.
e. Verify that this tangent passes through a control point.

Part 3 Other shapes

Draw Bezier curves and record the equation(s) for the following

a. a semicircle of radius 3.
b. the letter C in this font.
c. the letter S.

EXTENSION

Create a drawing of a profile of a car, using a combination of Bezier curves. (Bezier developed these curves for graphic design work in the car industry.)

LAB #51 Bezier curves

In this activity students

- Construct the locus using the Point Proportional tool
- Develop an understanding of the curve
- Drag endpoints to match a given curve
- Find parametric equations for the curve.

This lab demonstrates more involved mathematics, going beyond what most courses would require.

- Part 1 creates the curve. Generating and recording the equations could be omitted.
- Part 2 uses the control points to design a specific curve. The questions involve calculus of parametric equations and well developed algebraic skills.
- Part 3 is a further exploration of using the control points to generate specific shapes. You may wish to encourage students to be creative and design some shapes using a combination of Bezier curves.

Bezier curves are used extensively in design. Most drawing programs offer a curve drawing tool of this sort and the internal workings of the program is likely to be based on equations of spline curves. Students may have used some of these tools in programs that create vector graphics.

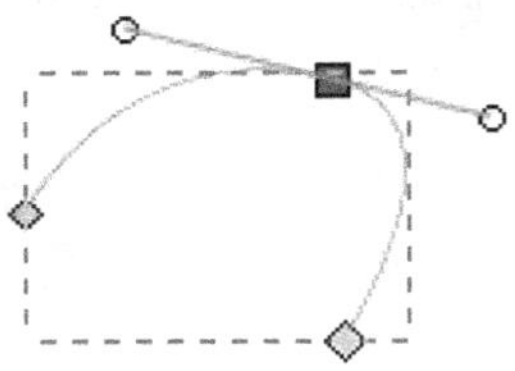

The picture shows such a curve drawn with the free open source program *Inkscape*. It clearly shows the end points and control points.

Looking at the equations should make it clear why it is called a cubic Bezier curve. It would be an interesting further investigation to use a different number of control points.
This Lab used 2 control points in addition to the start and end. If we used one what would the equations look like? What if we used three?

History
Pierre Bezier worked at Renault in the late 1960s. He was working with their early CAD/CAM software and wanted a simple way to specify smoothly bending curves. The result was Bezier curves. (Interestingly, Paul de Casteljau at Citroen had already developed exactly the same curves, but he was not allowed to publish by his company so they now bear Bezier's name)

Solution

Part 2 Answers will vary.

a. For end points (0, 0) and (30, 0) control points (–7.5, 13) and (37.5, 13) the equation is $\begin{matrix} x = -22.5t + 157.5t^2 - 105t^3 \\ y = 39t - 39t^2 \end{matrix}$ with $\frac{dx}{dt} = -22.5 + 315t - 315t^2, \frac{dy}{dt} = 39 - 78t$

b. Width extremeties occur when $\frac{dx}{dt} = 0$ i.e. $t = 0.923, 0.077$ and width = 31.7

c. Slope at P $= \frac{-39}{22.5}$ ($t = 0$), Q has $x = 0$ i.e. $t = 0.16$ and slope at Q = 1.34

d. $39x + 22.5y = 0$

e. $39x + 22.5y = 0$ passes through (–7.5, 13). Verified by substitution.

References

Fetter A. (1995) *Tethering my Thanksgiving Turkey,* Math Forum http://mathforum.org/pow/solutio44.html accessed 2008-07-24

Frossinakis, T. and Sheppard, I. (2007) *Agents for Change or Supporters of the Status Quo? We must decide!* Workshop presentation NCTM Annual Meeting Atlanta, http://mysite.verizon.net/tomarias777/, accessed 2008-07-24

Kansky B. *Algebra in the 21st Century Curriculum (K-16)*, http://mathforum.org/kb/thread.jspa?threadID=1710052&messageID=6132159, accessed 2008-07-24

Kozulin *et. al.* (2003) *Vygotsky's Educational Theory in Cultural Context*, Cambridge University Press, New York

GRADES 9-12: Standard 8 - Geometry from an Algebraic Perspective, National Council of Teachers of Mathematics (NCTM) (1989), http://my.nctm.org/standards/previous/currevstds/9-12s8.htm, accessed 2008-08-24

USA Mathematics Talent Search, http://usamts.org/, accessed 2008-07-24

Katz *V.* (2007) *Algebra: Gateway to a Technological Future*, Mathematical Association of America

Appendix A – GX and Functions

The following extract illustrates the function drawing capability of GX. The full document, written by Tim Brown and Jim Wiechmann (2008) can be downloaded from http://geometryexpressions.com/downloads/lessons/Using%20a%20Constraint%20Approach%20in%20Teaching%20Mathematics.pdf - Lesson 5.

Vertical Translations part A

I. $y = x^2$

$y = x^2$ is the parent function for a family of functions, whose graphs are a shape called a parabola. Pay particular attention to the vertex, as that is a key point for a parabola.

1) Draw the graphs below and investigate the larger pattern of what happens to the function as you add a constant number to it, then answer the questions below. Clearly label and/or color-code your graphs.

   GX will help you if you use the Draw | Function tool, and type in $y = x^2 + c$. Be careful, you must use * to indicate multiplication and ^ to indicate an exponent. Then use the variables tool palette, highlight c, and type in the appropriate number in the box below. You can see how the graph changes for a wider range of a values by clicking on the curve and dragging up and down, or by highlighting a in the Variables tool palette and dragging the scroll bar.

   A) $y = x^2$ B) $y = x^2 + 3$

   C) $y = x^2 + 5$ D) $y = x^2 - 2$

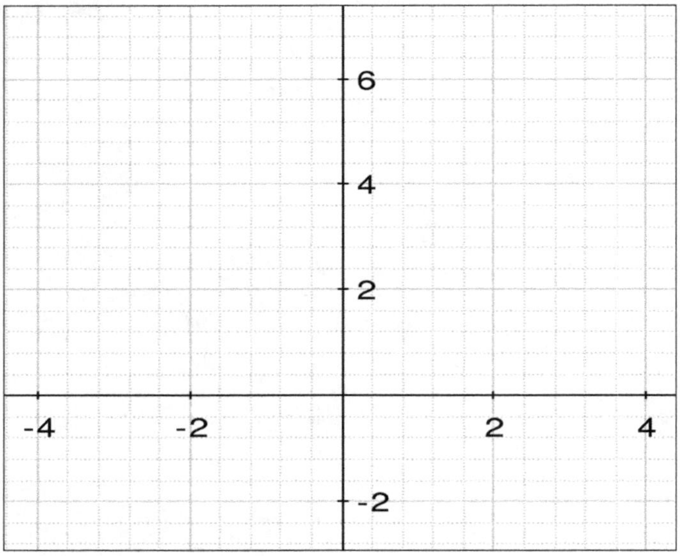

2) What seems to happen to the graph of the function when a constant number is added to it? Be sure to include the case where the number is negative.

Solution:

1) 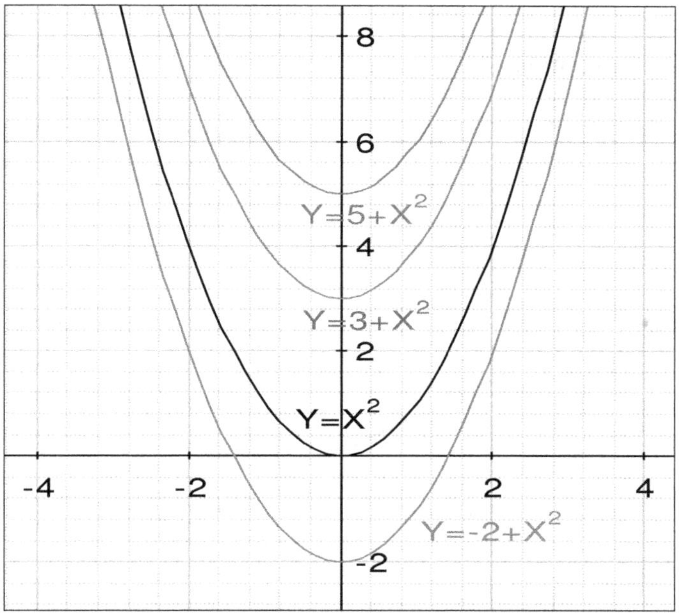

2) Answers may vary slightly. The graph is moving up (vertical translation) by the number of units that is being added to the function. If it is a negative number, the graph is moving down.

Note:

Using symbolics and the variable window enables the graph to be moved using the slider in the variable window and to animate the graph by varying one parameter at a time.

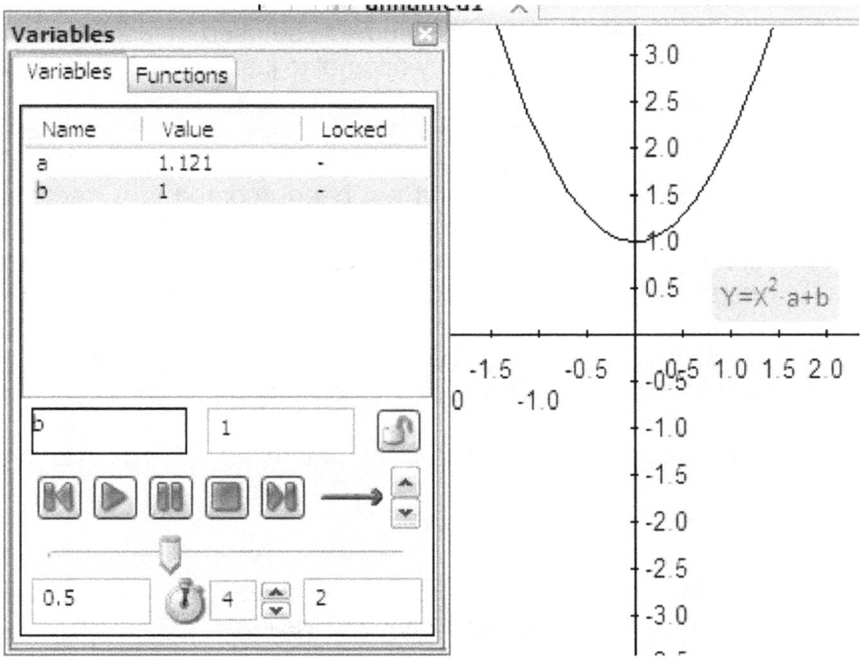

Appendix B - Insight with Geometry Expressions

Introduction

Geometry Expressions automatically generates algebraic expressions from geometric figures. For example in the diagram below, the user has specified that the triangle is right and has short sides length a and b. The system has calculated an expression for the length of the altitude:

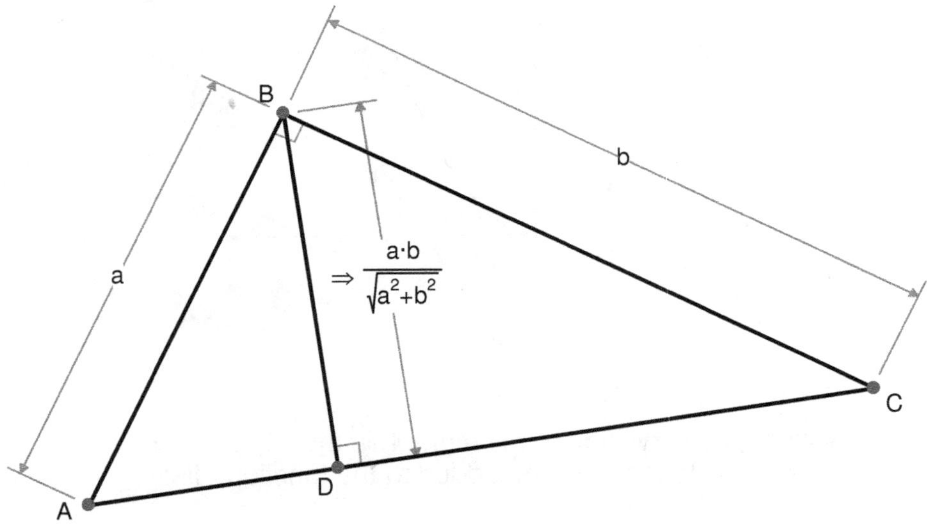

Can a system which does geometry be used in teaching geometry? I'd suggest the answer is yes. And this article is my attempt to justify that answer by example.

In 1981, I was supplementing my income as a research assistant by tutoring mathematics. One of my students was studying for his "City and Guild" exam in electronics. The examination's contents had not been updated in a decade, and part of the exam was to perform multiplication and power calculations using log tables. Mathematicians of my age or older will remember the peculiar arithmetic used with log tables where numbers less than 1 were involved. By 1982 the advent of inexpensive pocket calculators had completely eliminated the need to use log tables. However the exam still contained these problems, and my student had to learn how to do them, except instead of looking up log and antilog tables, he used the log function on his calculator.

Technology, I contend, should be embraced rather than ignored.

Warm Up

The main section of this article is an investigation of a specific geometric topic (Incircles and Circumcircles) using Geometry Expressions.

As a warm up, we'll examine a handful of simple examples in which we hope to show how Geometry Expressions can be as part of a process of creative mathematics.

A sequence of altitudes

We're going to look at the figure we showed in the introduction:

As this is Example 1, I'll show you in some detail how to create this diagram:

We will use three toolbars, the Draw toolbar to create the geometry, the Constrain toolbar to specify lengths and angles, and the Calculate toolbar to measure the length of the altitude.

First however, we sketch the figure using the line segment tool:

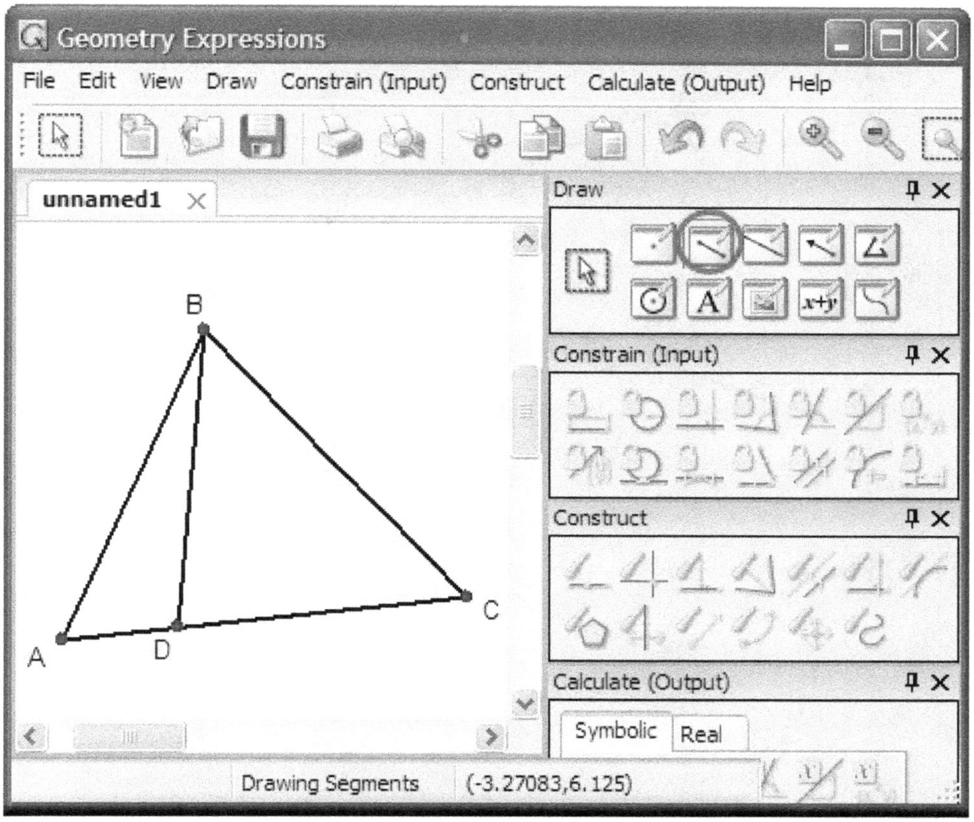

Next, we change to select mode, to add the right angle constraints and the length constraints for lines AB and CD:

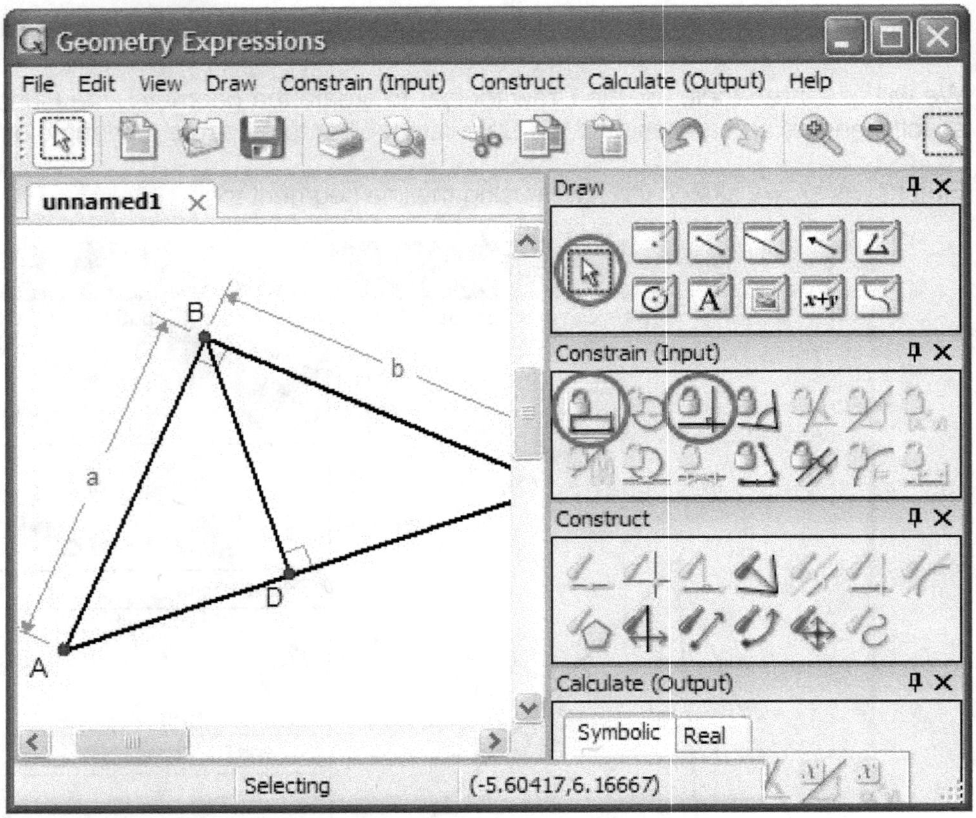

Finally, calculate the length of BD, by selecting it then clicking the Calculate Symbolic Length button:

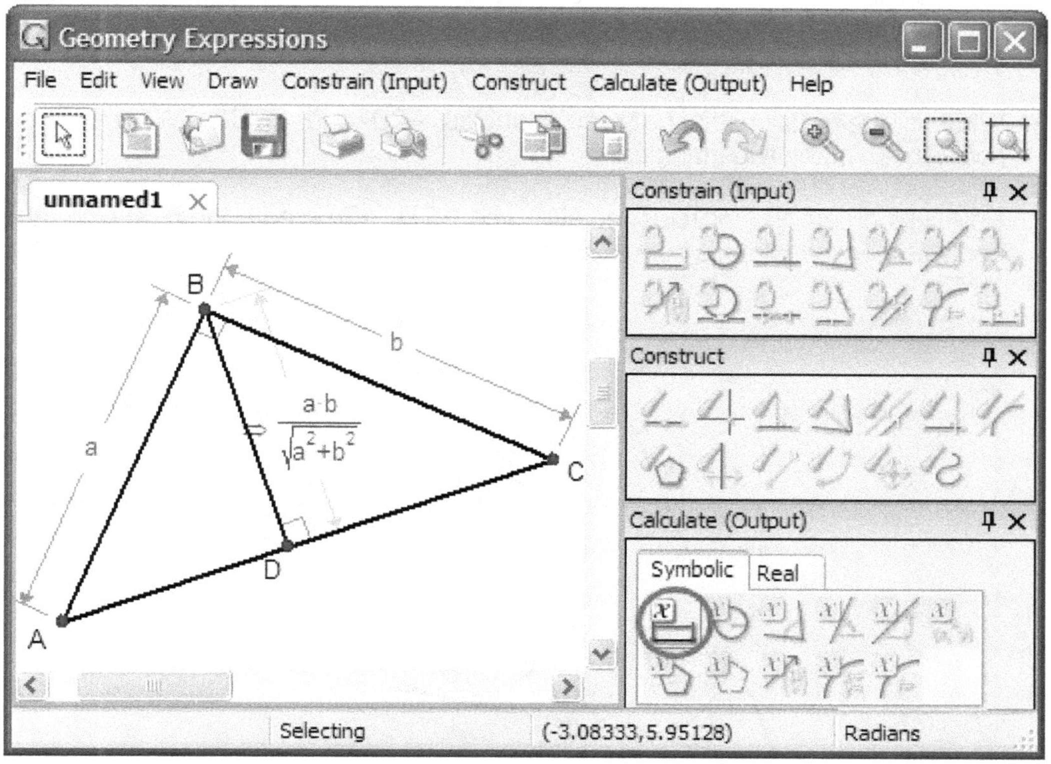

Can you prove that the expression for the altitude is correct? (Think about the relationship between altitude and area for a triangle).

Now let's look at the length AD:

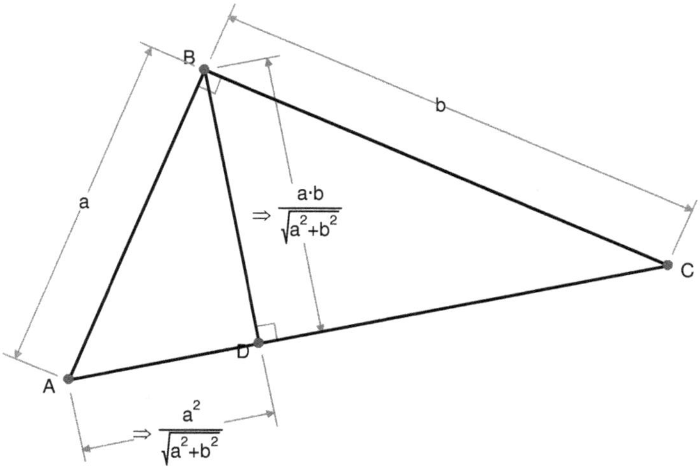

What is the ratio |AD| / |BD|? What does this tell you about the relationship between the triangle ABD and the triangle ABC? Can you establish this relationship in a different way (think angles)? Hence, can you prove the formula for AD?

What is the length CD? What is the ratio CD / BD?

What is the relationship between triangles ABD and BCD? What is the ratio of the hypotenuse of the two triangles?

Now let's create the incircles for ABD and BCD. (To create them, sketch the circles, then apply tangent constraints between the circles and the appropriate sides of the triangles)

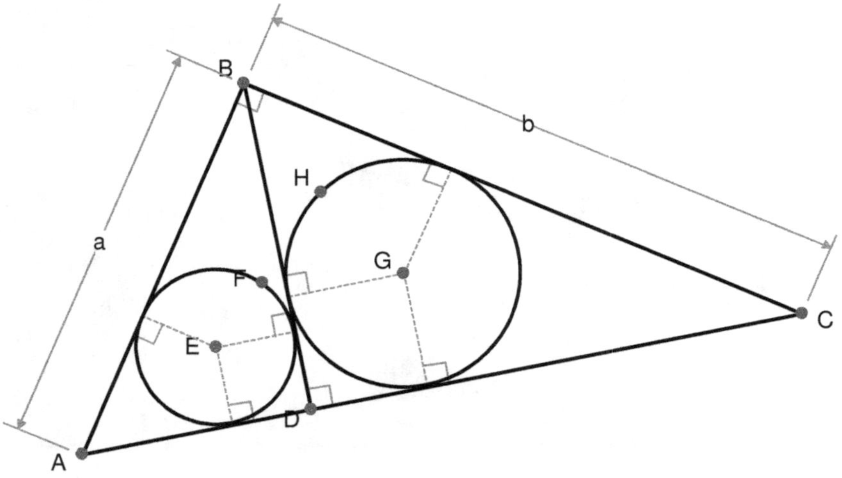

What is the ratio of the radii of these incircles?

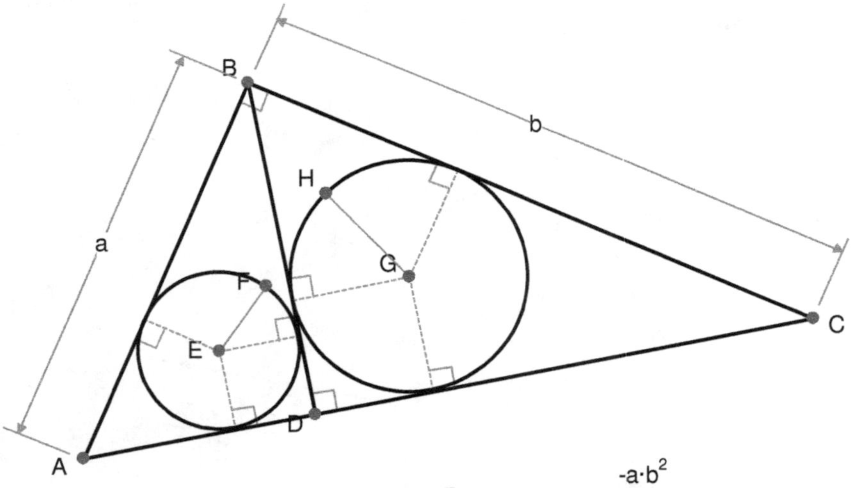

$$z_5 \Rightarrow \frac{-a\cdot b^2}{-b^2+a\cdot\left(-a\cdot\sqrt{a^2+b^2}\right)-b\cdot\sqrt{a^2+b^2}}$$

$$z_4 \Rightarrow \frac{a^2\cdot b}{a^2+b^2+a\cdot\sqrt{a^2+b^2}+b\cdot\sqrt{a^2+b^2}}$$

$$\frac{z_4}{z_5} \Rightarrow \frac{a}{b}$$

We see the ratio is a/b. Is this a surprise?

Let's go back to our original drawing and create another altitude:

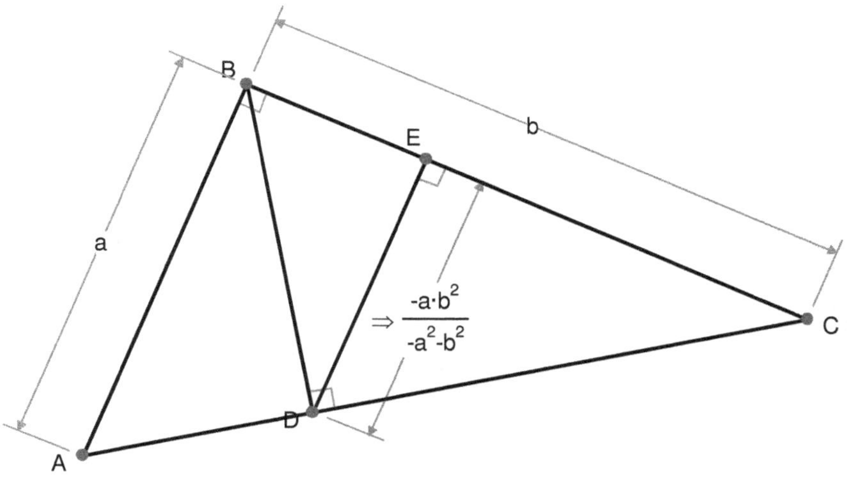

What is the ratio |DE|/|AB|? Can you prove the formula for |DE|?

Can you predict the length of FG in the drawing below?

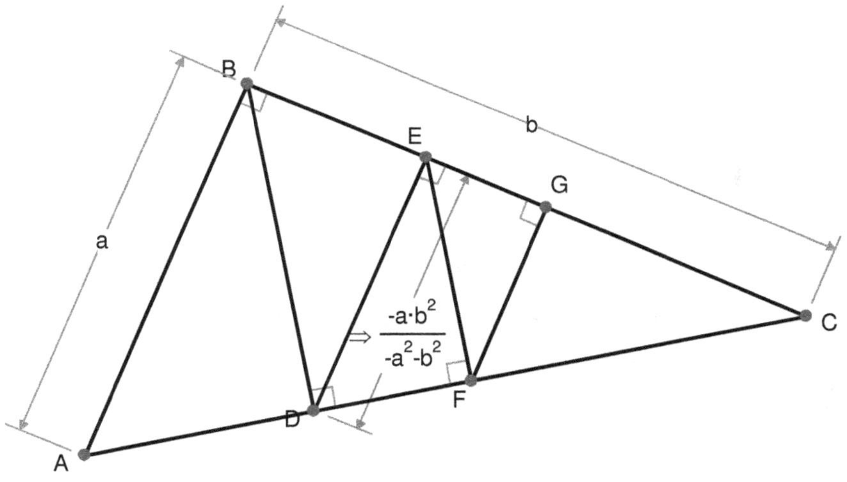

Angles and Circles

If a chord subtends an angle of θ at the center of a circle what does it subtend at the circumference?

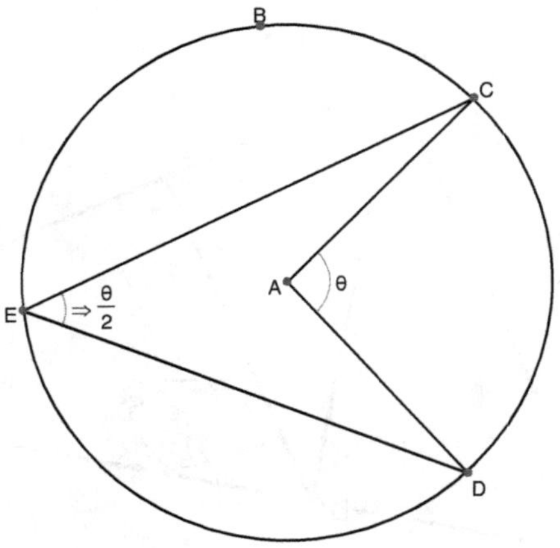

Can you prove this result? (Hint: start with the diagram below and fill in the angles).

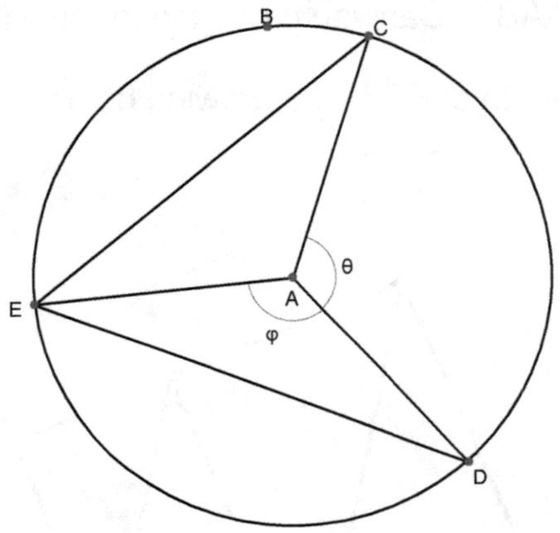

Here's a sequence of diagrams, does this constitute a proof?

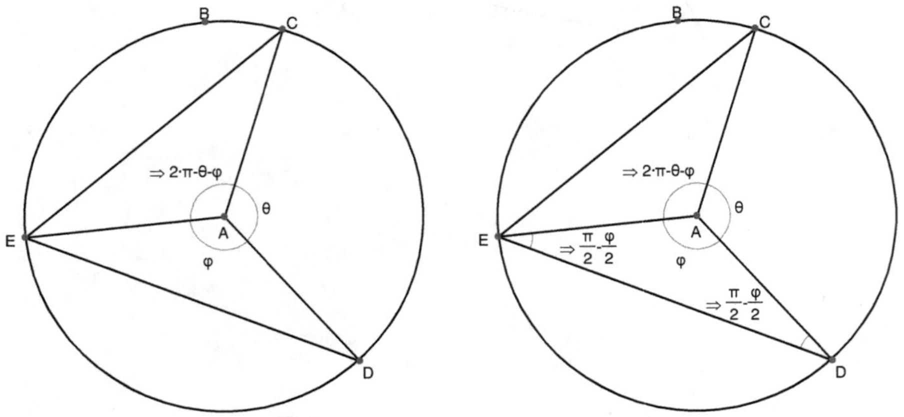

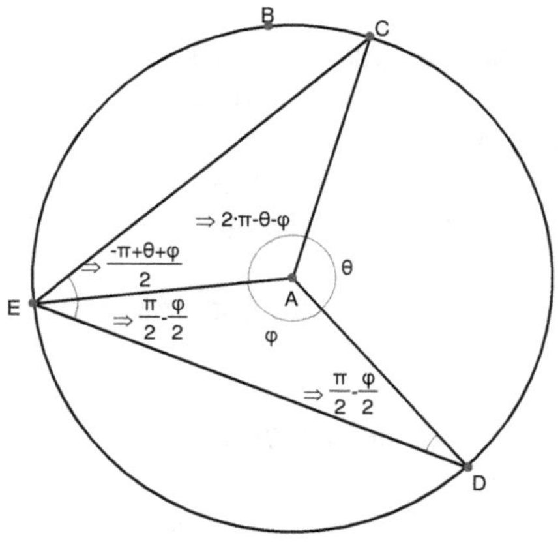

Can you follow a similar approach and prove these results:

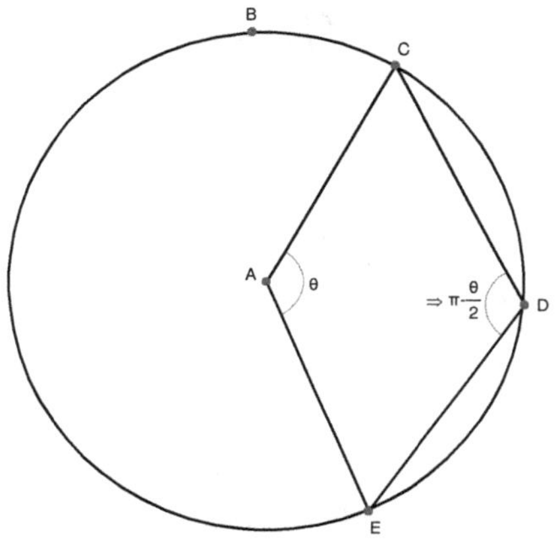

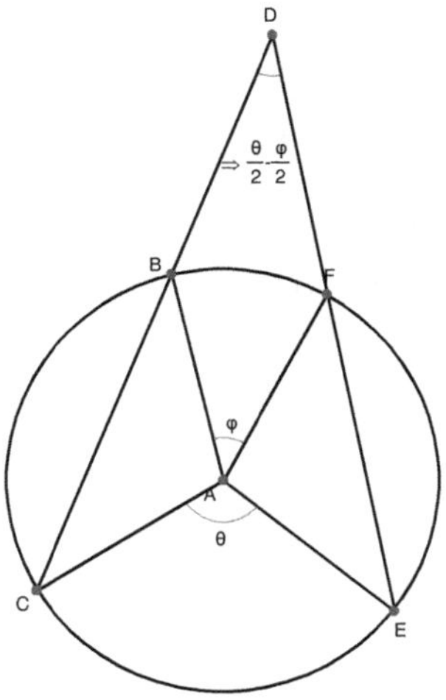

Triangulation

Geometry Expressions has its own algebra system customized to manipulating the sort of mathematics that arises in geometry problems. However, to do further analysis of your geometry expressions you should copy them into a more fully featured algebra system. Geometry Expressions exports expressions in the MathML format, which is accepted as input by a wide variety of mathematics display and computation applications.

In this exercise, we will work on an example which involves copying into an algebra system to complete the analysis.

Assume we are writing a computer program which performs triangulation: given the length of a baseline and angles measured off the baseline to the apex of the triangle, the program is to give the coordinates of the apex. The appropriate expressions can be derived from Geometry Expressions:

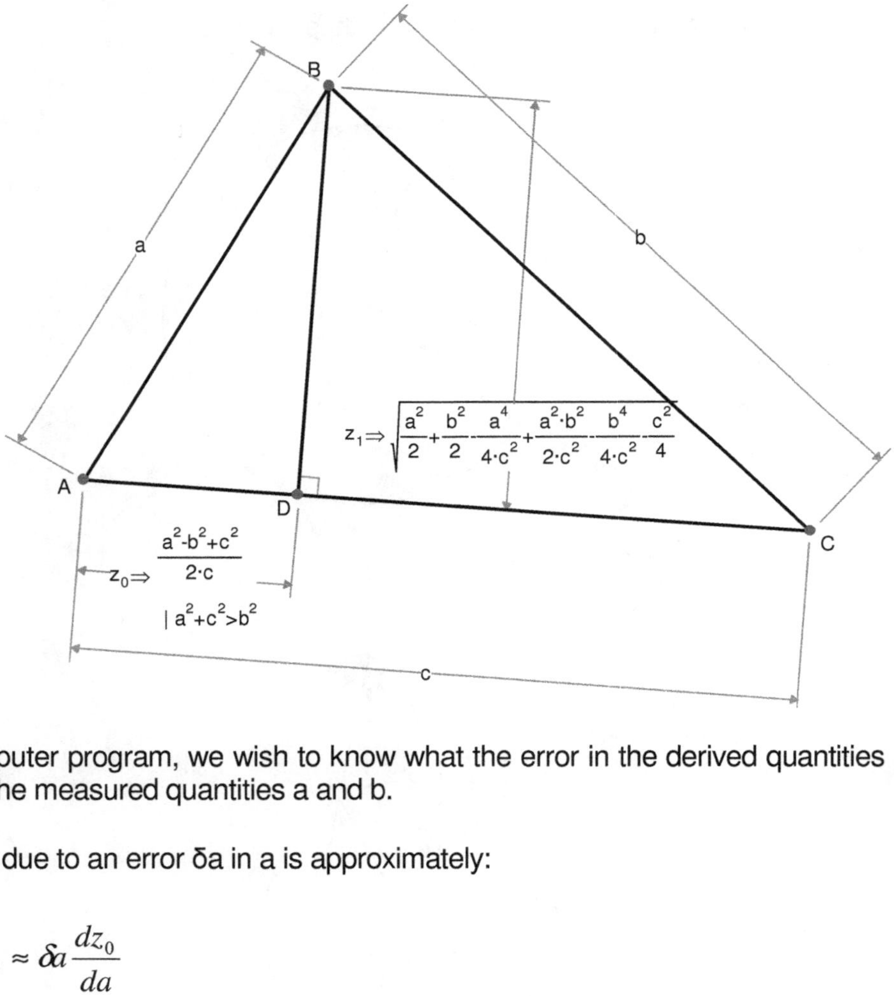

In the computer program, we wish to know what the error in the derived quantities is relative to error in the measured quantities a and b.

Error in z_0 due to an error δa in a is approximately:

$$\delta z_0 \approx \delta a \frac{dz_0}{da}$$

And similarly for error in z_1

The quantity $\frac{dz_0}{da}$ can be thought of as the error magnifier.

If you select the expression z_0 you can copy and paste into an algebra system: Maple, or Mathematica or any other system which is prepared to accept mathML. We'll use the Casio ClassPad Manager. Once in there, we'll differentiate with respect to a to get an expression for the error magnification:

$$\frac{d}{da}\left(\frac{a^2+b^2\cdot(-1)+c^2}{c\cdot 2}\right)$$

$$\frac{a}{c}$$

$$\frac{d}{da}\left(\sqrt{\frac{a^2}{2}+\frac{b^2}{2}+\frac{a^4\cdot(-1)}{c^2\cdot 4}+\frac{b^2\cdot a^2}{c^2\cdot 2}+\frac{b^4\cdot(-1)}{c^2\cdot 4}+c^2\cdot\left(-\frac{1}{4}\right)}\right)$$

$$\frac{-\left(a^3-a\cdot b^2-a\cdot c^2\right)}{|c|\cdot\sqrt{-a^4-b^4-c^4+2\cdot a^2\cdot b^2+c^2\cdot\left(2\cdot a^2+2\cdot b^2\right)}}$$

The error magnifier in z_0 is simply the ratio of a/c. In z_1 the magnifier is more complicated. Factoring the term under the square root gives us a clearer picture:

$$\text{factor}\left(-a^4-b^4-c^4+2\cdot a^2\cdot b^2+c^2\cdot\left(2\cdot a^2+2\cdot b^2\right)\right)$$

$$-(a+b+c)\cdot(a-b+c)\cdot(a-b-c)\cdot(a+b-c)$$

We see that the denominator of the error term goes to zero when a=b+c, or when b=a+c, or when c=a+b.

Can you interpret these conditions geometrically?

Can you interpret the complete error term geometrically? Hint: compare the denominator with the area of the triangle. Compare the numerator with the distance AD in the diagram. Can you construct a distance on the diagram whose length is more closely related to the numerator?

One question we might ask is this: what is the optimal geometry for triangulating? Let's simplify the question by assuming the triangle is isosceles and a=b=x. We'll also assume the base length is 1.

$$\frac{d}{da}\left(\frac{a^2+b^2\cdot(-1)+c^2}{c\cdot 2}\right)\Big|\{a=x,b=x,c=1\}$$

$$x$$

$$\frac{d}{da}\left(\sqrt{\frac{a^2}{2}+\frac{b^2}{2}+\frac{a^4\cdot(-1)}{c^2\cdot 4}+\frac{b^2\cdot a^2}{c^2\cdot 2}+\frac{b^4\cdot(-1)}{c^2\cdot 4}+c^2\cdot\left(-\frac{1}{4}\right)}\right)\Big|\{a=x,b=x,c=1\}$$

$$\frac{x}{\sqrt{4\cdot x^2-1}}$$

We can graph these functions and observe that for large values of x the first dominates, for small values of x the second dominates.

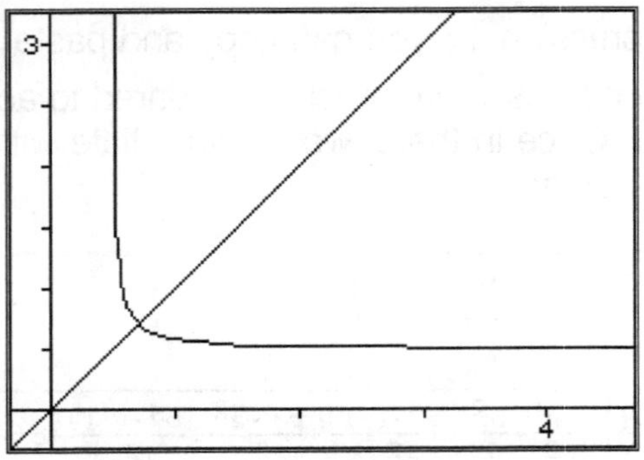

The optimal value of x is when the two are equal:

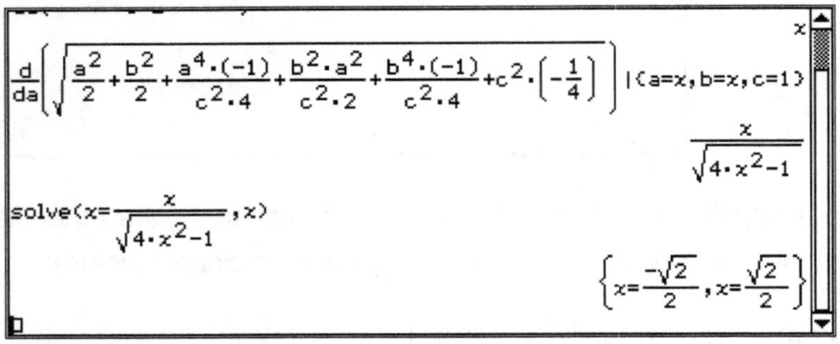

What angles does this triangle have?

Rectangle Circumscribing an Equilateral Triangle

In page 19-21 of Mathematical Gems, by Ross Honsberger (and various other places), we have the following theorem: Inscribe an equilateral triangle in a square such that one corner of the triangle is a corner of the square and the other two corners lie on the opposite sides of the square. This forms 3 right triangles. The theorem states:

The area of the larger right triangle is the sum of the areas of the smaller two.

You can measure areas in *Geometry Expressions* by selecting a connected set of segments and constructing a polygon. You can then create an Area measure for that polygon.

The diagram below shows the areas of the triangles in the above theorem

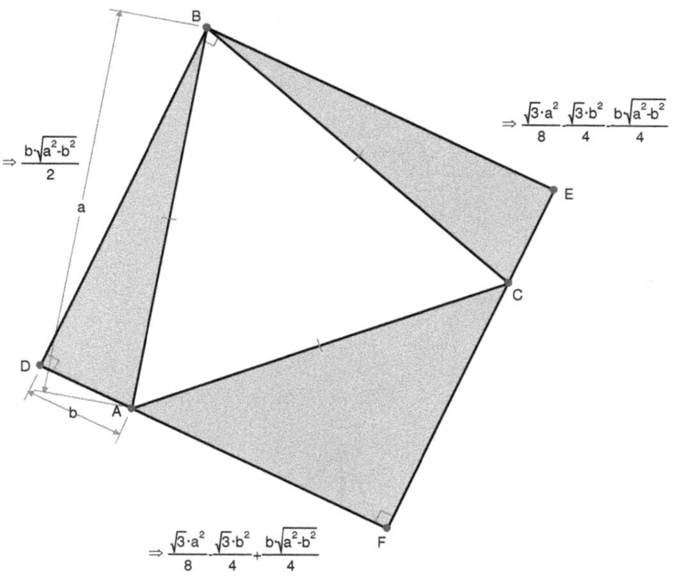

Can you prove the theorem from the diagram?

Can you prove that the diagram is correct?

Area of a Hexagon bounded by Triangle side trisectors

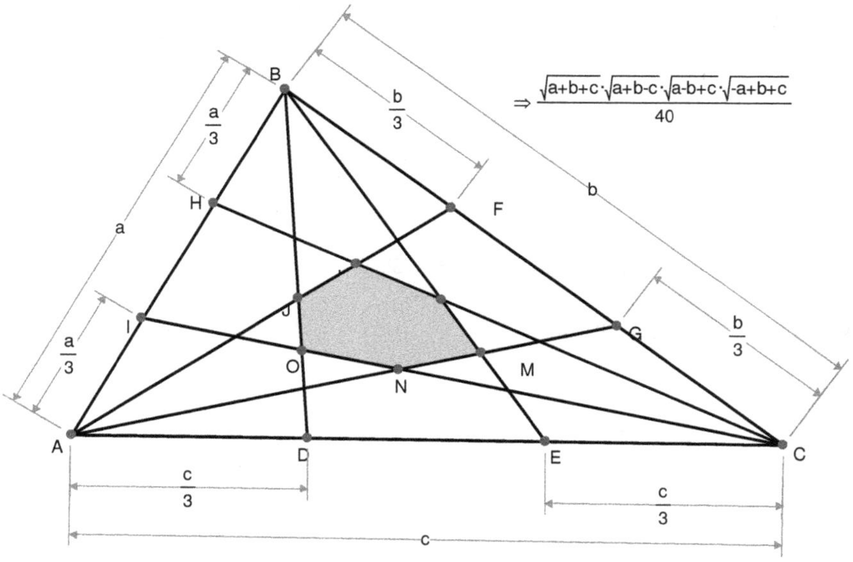

The shaded hexagon is formed by intersecting the lines joining the vertices of the triangle with the trisectors of the opposite sides.

How does the area relate to the area of the triangle ABC?

Can you prove this?

One approach uses this expression for the location of the intersection of CE and BD where E is proportion t along AB and D is proportion t along AC. Use values 1/3 and 2/3 for t to get points of the hexagon.

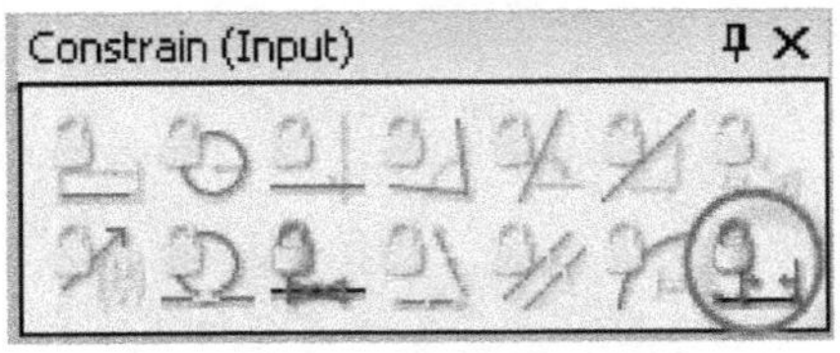

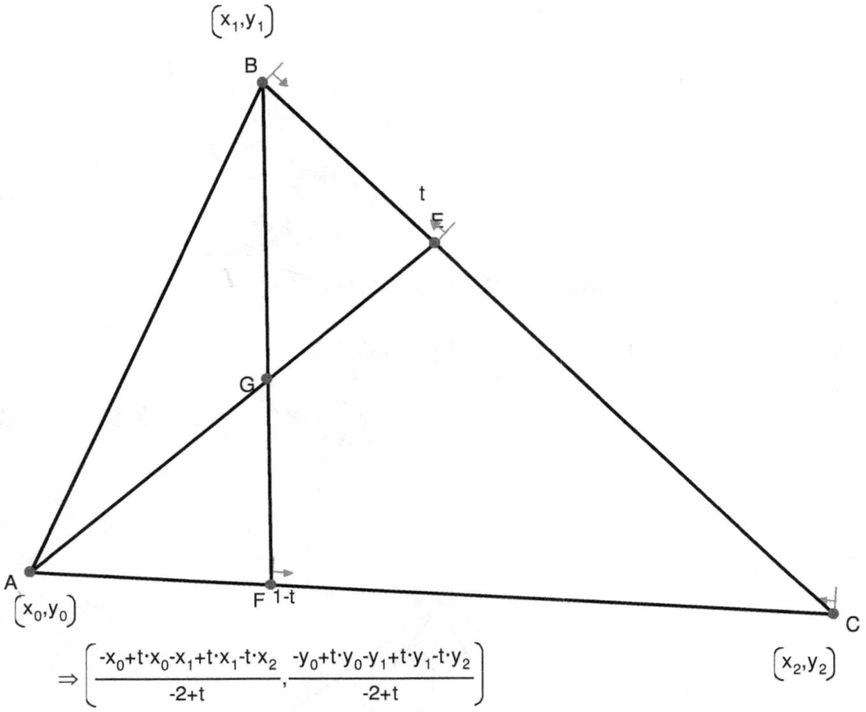

Can you prove the area of the hexagon using this information?

Can you prove the coordinates?

How about the area of the other hexagon?

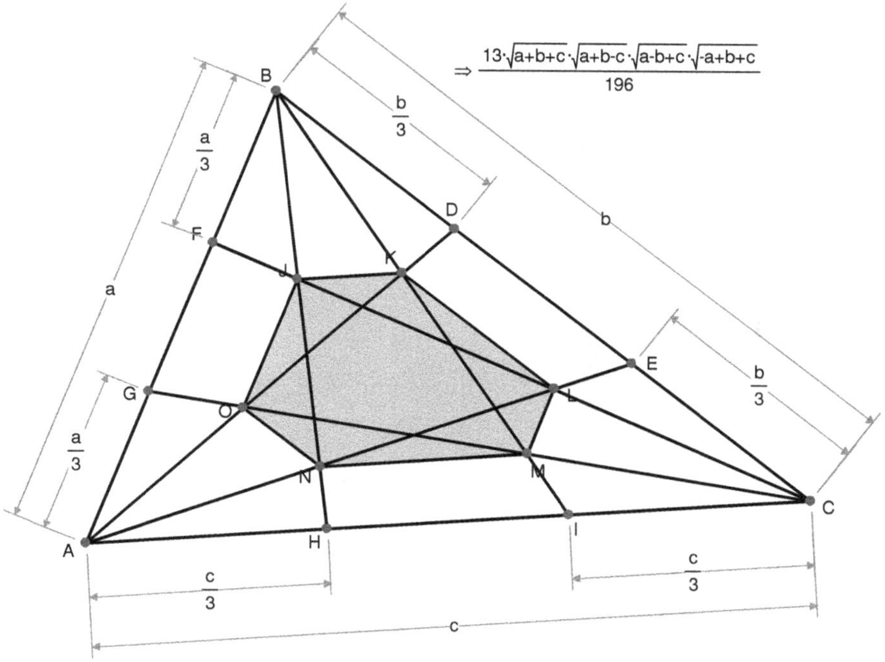

An Investigation of Incircles, Circumcircles and related matters

We'll follow with a set of examples all related to the theme of incircles, circumcircles, excircles and triangle areas.

Circumcircle Radius

Measure the area of a triangle sides length a,b,c, and measure the radius of the circumcircle.

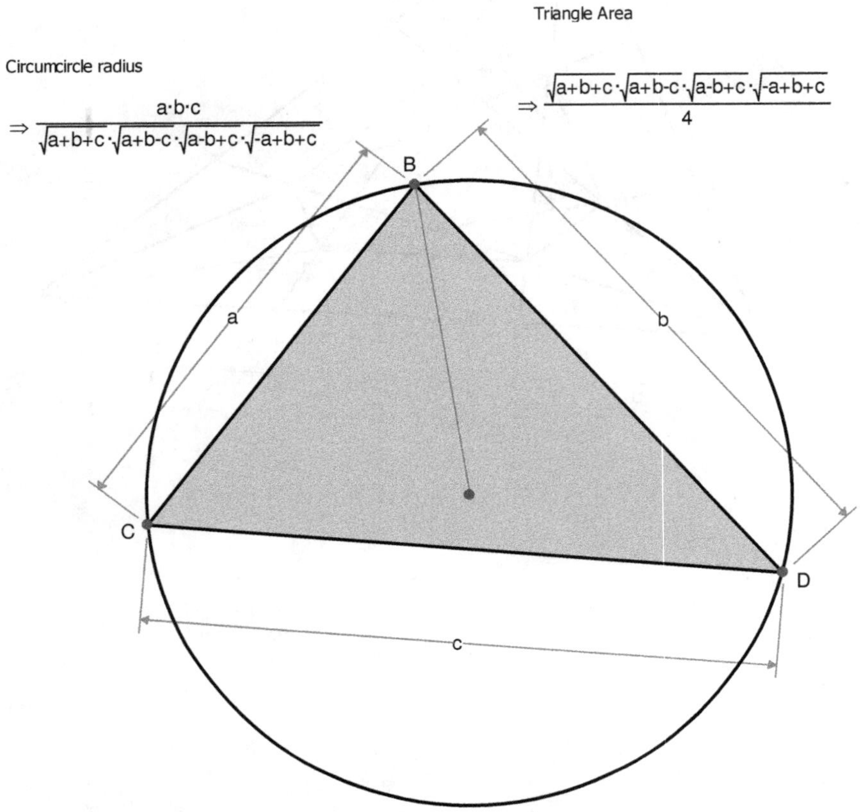

What is their relationship?

If a triangle is defined in terms of two sides and the included angle, what is its area?

Hence, derive a formula for the radius of the circumcircle which involves an angle:

You can see this in *Geometry Expressions*:

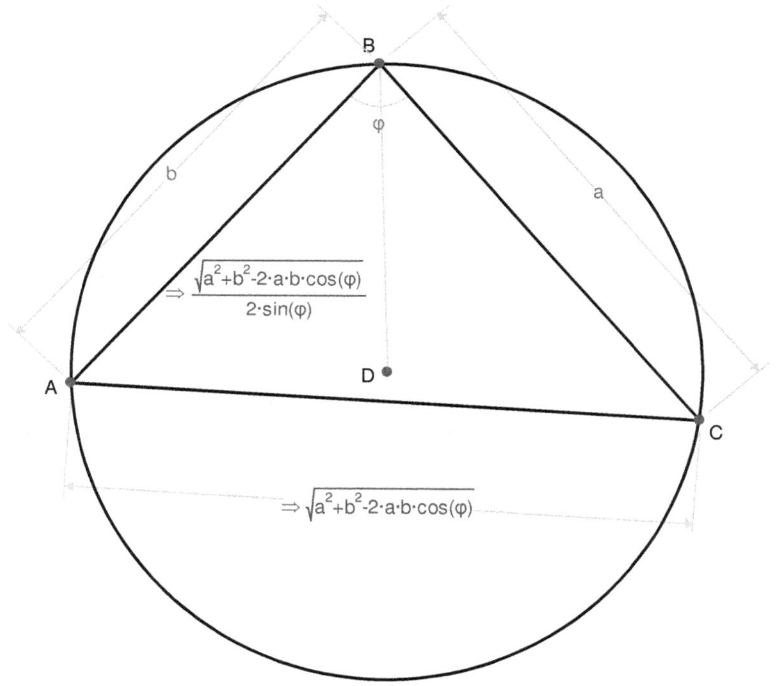

Can you prove that this expression is true independent of Geometry Expressions? Hint, in the diagram below, get an expression for |BC|

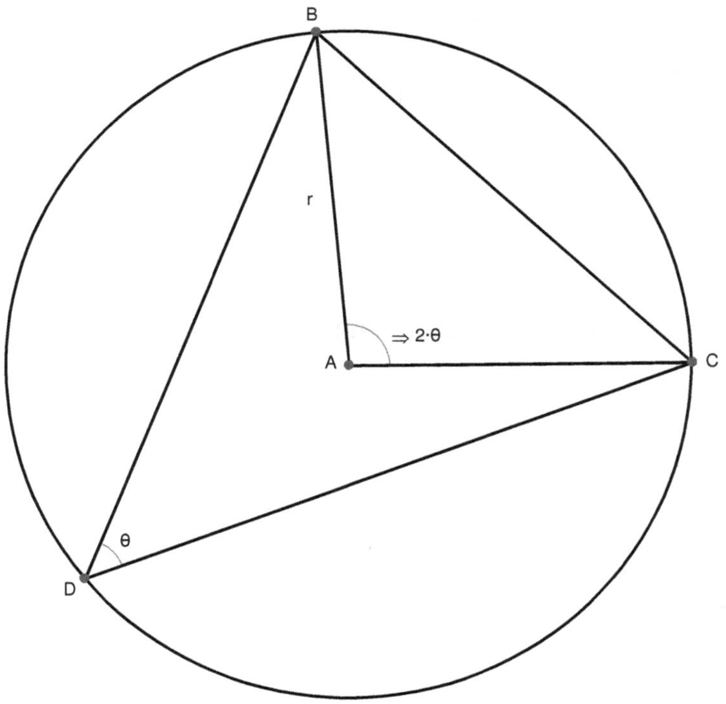

Let R be the radius of the circumcircle, let A,B,C be the angles of a triangle whose opposite sides have length a,b,c. At this point, you should have proved:

$$R = \frac{c}{2\sin(C)}$$

$$Area(ABC) = \frac{1}{2}ab\sin(C) = \frac{abc}{4R}$$

Hence:

$$R = \frac{abc}{4Area(ABC)}$$

Now to prove the original formula for R in terms of a,b,c, we need only prove (or accept without proof) the formula for the area of the triangle generated by Geometry Expressions.

We don't like to accept anything without proof do we?

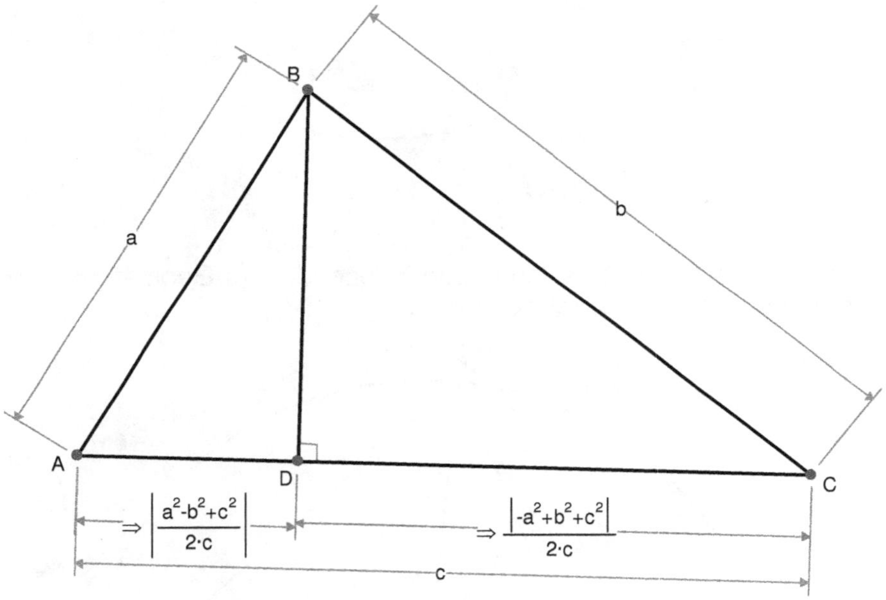

Show that with the above values for |AD| and |CD|,

$$a^2 - |AD|^2 = b^2 - |CD|^2$$
$$|AD| + |DC| = c$$

Copy the expression for AD into your algebra system. Now create an expression for the square of the altitude. Multiply by the square of c, and divide by 4 to create an expression for the square of the area:

```
> c^2*(a^2-(1/2*(a^2-b^2+c^2)/c)^2)/4;
```

$$\frac{1}{4}c^2\left(a^2-\frac{1}{4}\frac{(a^2-b^2+c^2)^2}{c^2}\right)$$

You can then do some algebraic manipulation to get this into a nicer form:

```
> expand(%);
```

$$\frac{1}{8}a^2c^2-\frac{1}{16}a^4+\frac{1}{8}a^2b^2-\frac{1}{16}b^4+\frac{1}{8}b^2c^2-\frac{1}{16}c^4$$

```
> simplify(%);
```

$$\frac{1}{8}a^2c^2-\frac{1}{16}a^4+\frac{1}{8}a^2b^2-\frac{1}{16}b^4+\frac{1}{8}b^2c^2-\frac{1}{16}c^4$$

```
> factor(%);
```

$$-\frac{1}{16}(b+a+c)(b+a-c)(-c+a-b)(a-b+c)$$

As a final exercise, let's look at the center of the circumcircle in terms of the coordinates of the triangle vertices (barycentric coordinates)

$$\Rightarrow\left(\frac{(y_1-y_2)\cdot\left(\frac{(x_0+x_1)\cdot(x_0-x_1)}{2}+\frac{(y_0+y_1)\cdot(y_0-y_1)}{2}\right)+(y_0-y_1)\cdot\left(\frac{(x_1+x_2)\cdot(-x_1+x_2)}{2}+\frac{(y_1+y_2)\cdot(-y_1+y_2)}{2}\right)}{-x_1\cdot y_0+x_2\cdot y_0+x_0\cdot y_1-x_2\cdot y_1-x_0\cdot y_2+x_1\cdot y_2},\frac{-(x_1-x_2)\cdot\left(\frac{(x_0+x_1)\cdot}{2}\right.}{}\right.$$

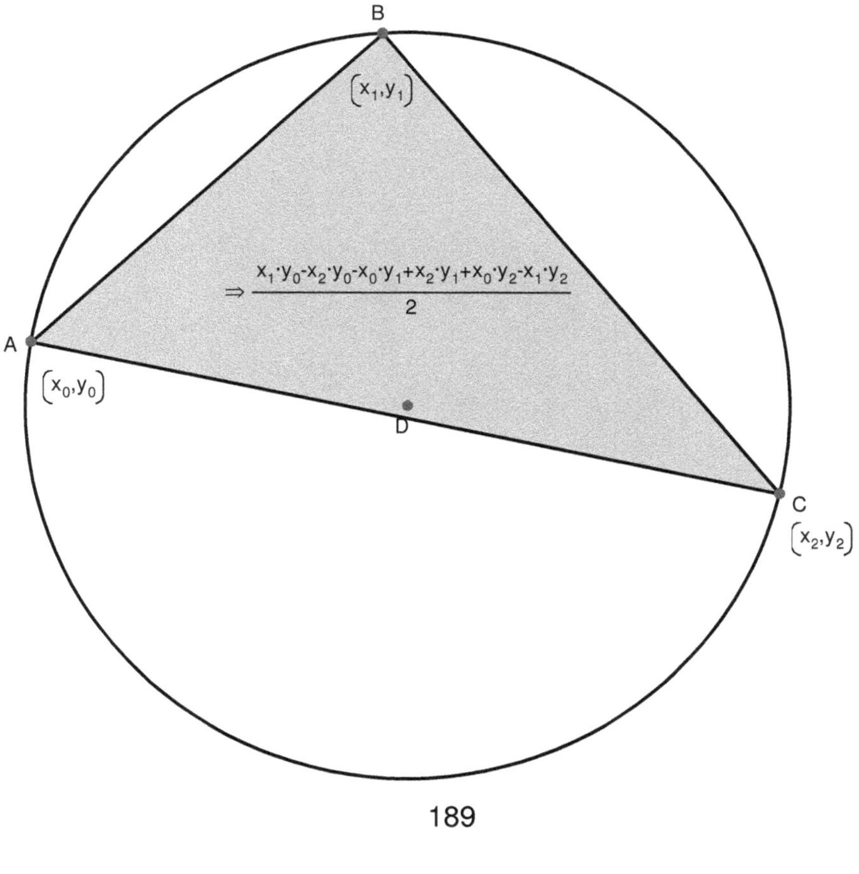

The expression is quite complicated, but breaks down into constituent parts. Do you see the area embedded in the formula?

Can you write this simpler in vector terms?

$$\Rightarrow \left(\frac{-u_1^2 \cdot v_0 - v_0 \cdot v_1^2 + v_1 \cdot \left(u_0^2 + v_0^2 \right)}{2 \cdot \left(-u_1 \cdot v_0 + u_0 \cdot v_1 \right)}, \frac{-u_0^2 \cdot u_1 + u_0 \cdot u_1^2 - u_1 \cdot v_0^2 + u_0 \cdot v_1^2}{2 \cdot \left(-u_1 \cdot v_0 + u_0 \cdot v_1 \right)} \right)$$

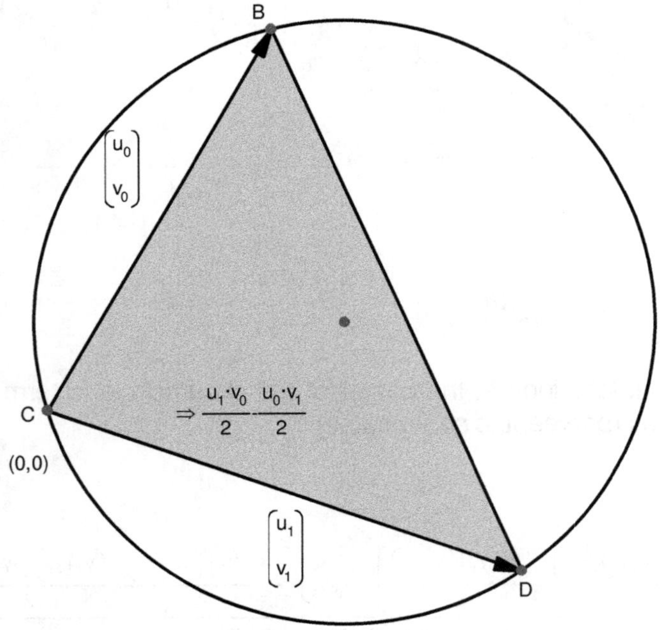

Incircle Radius

Here is the formula for the incircle radius (along with the familiar formula for the area of the triangle).

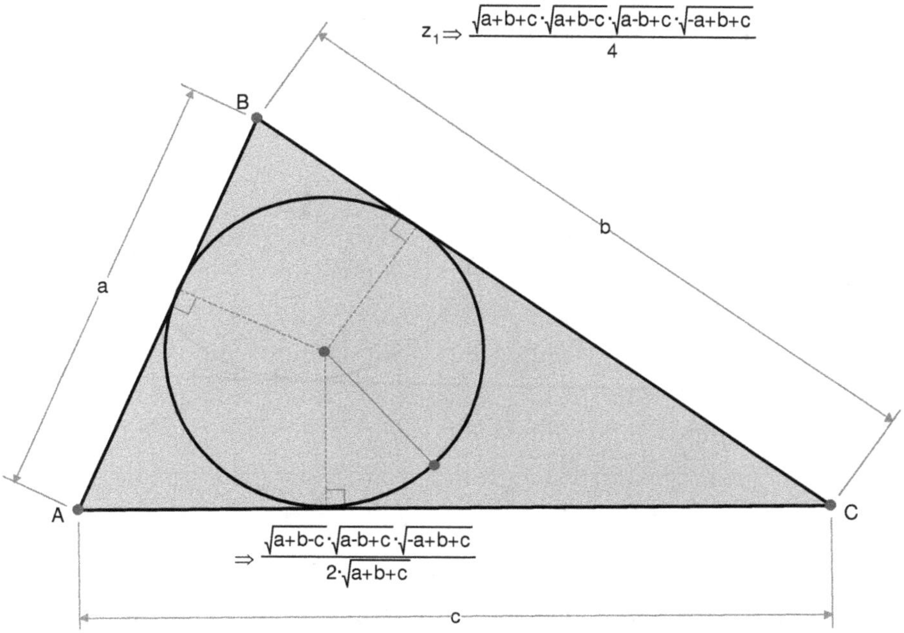

Can you express the incircle radius in terms of the area?

Now can you prove it independently of Geometry Expressions?

Hint: consider the areas of the shaded triangle:

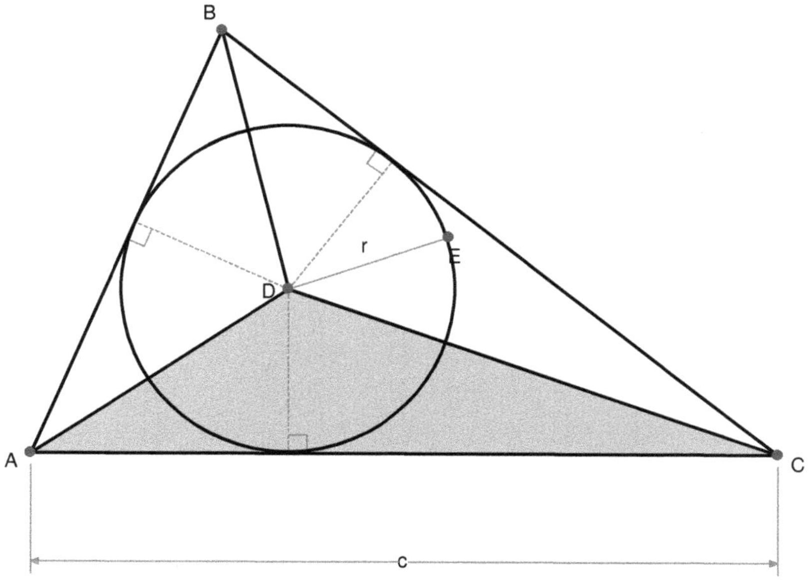

Incircle Center in Barycentric Coordinates

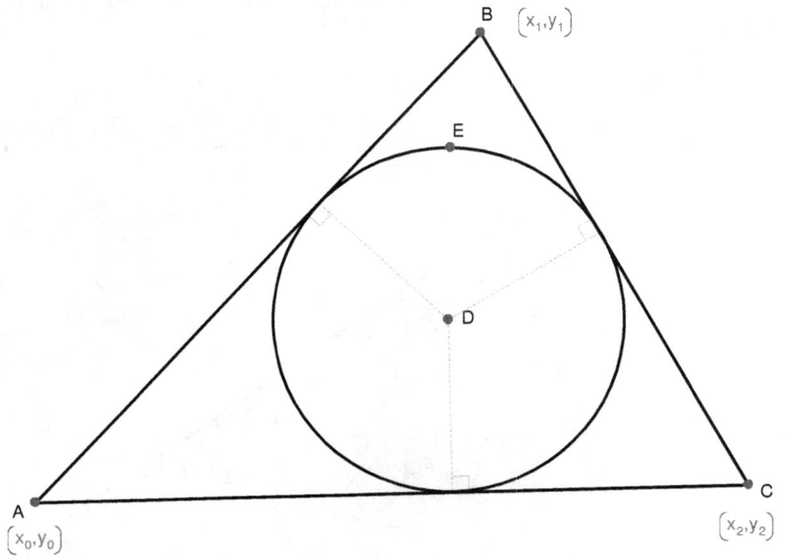

$$\Rightarrow \left(\frac{x_1\cdot\sqrt{(x_0-x_2)^2+(-y_0+y_2)^2}+x_2\cdot\sqrt{(-x_0+x_1)^2+(y_0-y_1)^2}+x_0\cdot\sqrt{(-x_1+x_2)^2+(y_1-y_2)^2}}{\sqrt{(x_0-x_2)^2+(-y_0+y_2)^2}+\sqrt{(-x_0+x_1)^2+(y_0-y_1)^2}+\sqrt{(-x_1+x_2)^2+(y_1-y_2)^2}}, \frac{y_1\cdot\sqrt{(x_0-x_2)^2+(-y_0+y_2)^2}+y_2\cdot\sqrt{(-x_0+x_1)^2+(y_0-y_1)^2}+y_0\cdot\sqrt{(-x_1+x_2)^2+(y_1-y_2)^2}}{\sqrt{(x_0-x_2)^2+(-y_0+y_2)^2}+\sqrt{(-x_0+x_1)^2+(y_0-y_1)^2}+\sqrt{(-x_1+x_2)^2+(y_1-y_2)^2}} \right)$$

If we let a = |BC|, b=|AC| and c=|AB| and let **A**, **B** and **C** be the position vectors of the points A,B,C then the incircle center is:

$$\frac{\mathbf{A}a+\mathbf{B}b+\mathbf{C}c}{a+b+c}$$

Create a point with barycentric coordinates (r,s,1-r-s) and examine the areas of the 3 triangles defined by the point and the three sides of the triangle. (You should lock the parameter values r and s to some values such that r+s<1, then the point will lie inside the triangle):

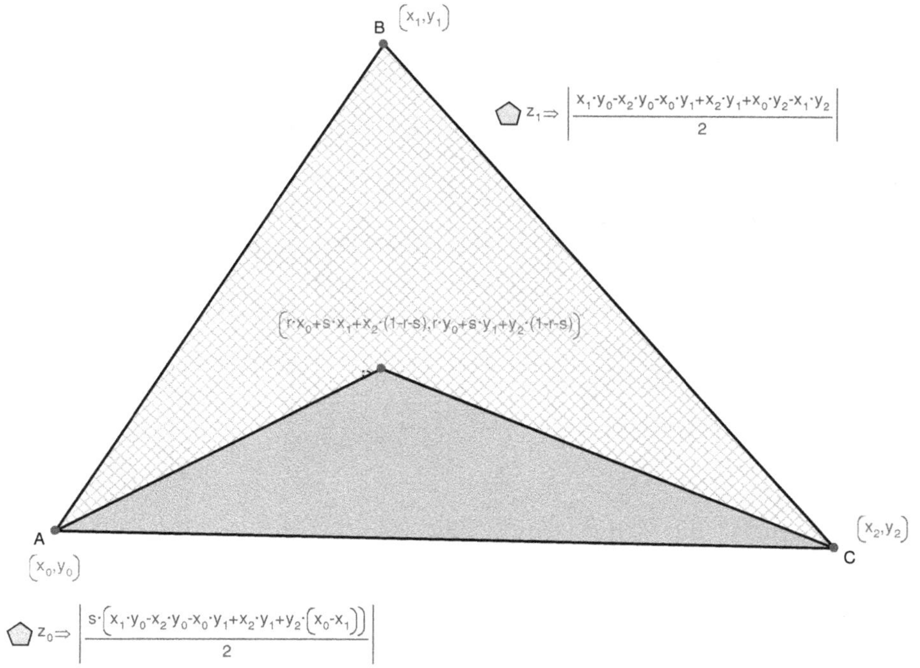

D has barycentric coordinates (r,s,1-r-s)
Note that the ratio of the areas ADC/ABC is the barycentric coordinate s

What is the ratio of this area to the area of the original triangle?

Can you use this relationship to prove the formula for the barycentric coordinates of the incenter?

Can you use this relationship to express the barycentric coordinates of the circumcenter in terms of the lengths of the sides of the triangle?

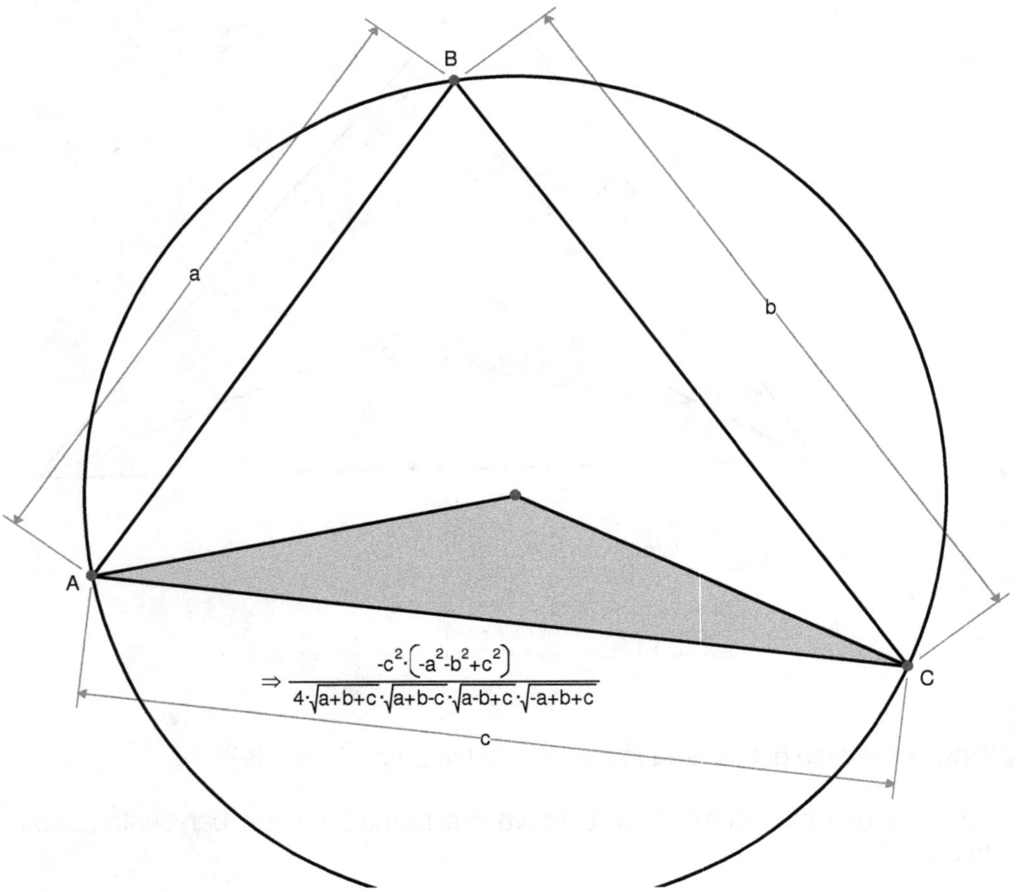

How does the point of contact with the incircle split a line?

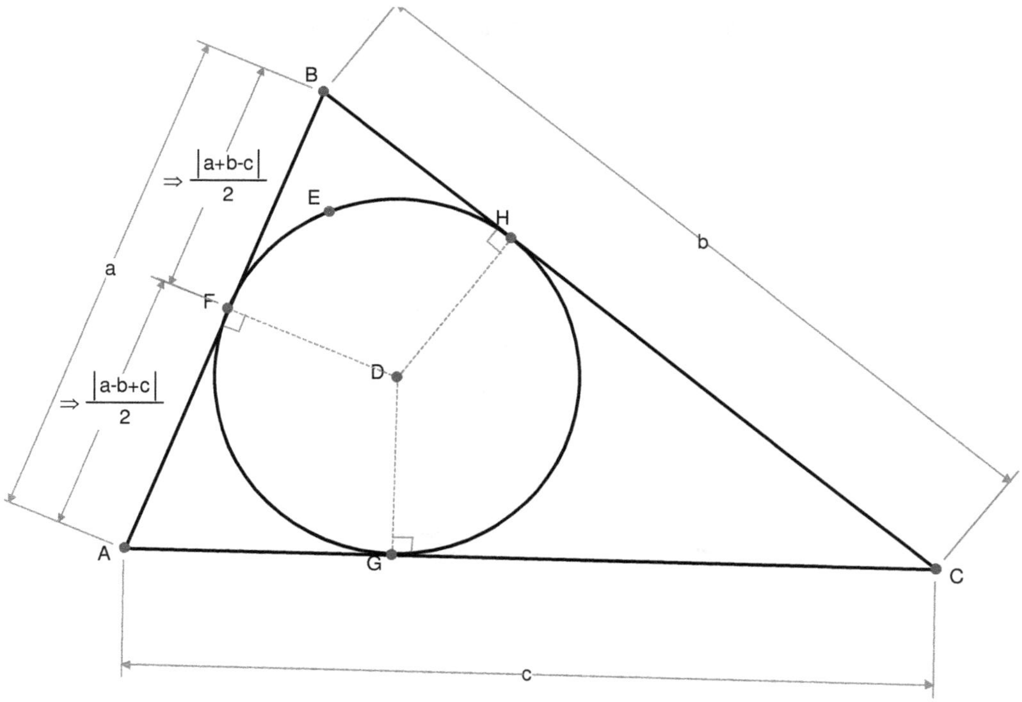

Can you deduce the lengths BH, HC, AG, GC?

Excircles

The three excircles of a triangle are tangent to the three sides but exterior to the circle:

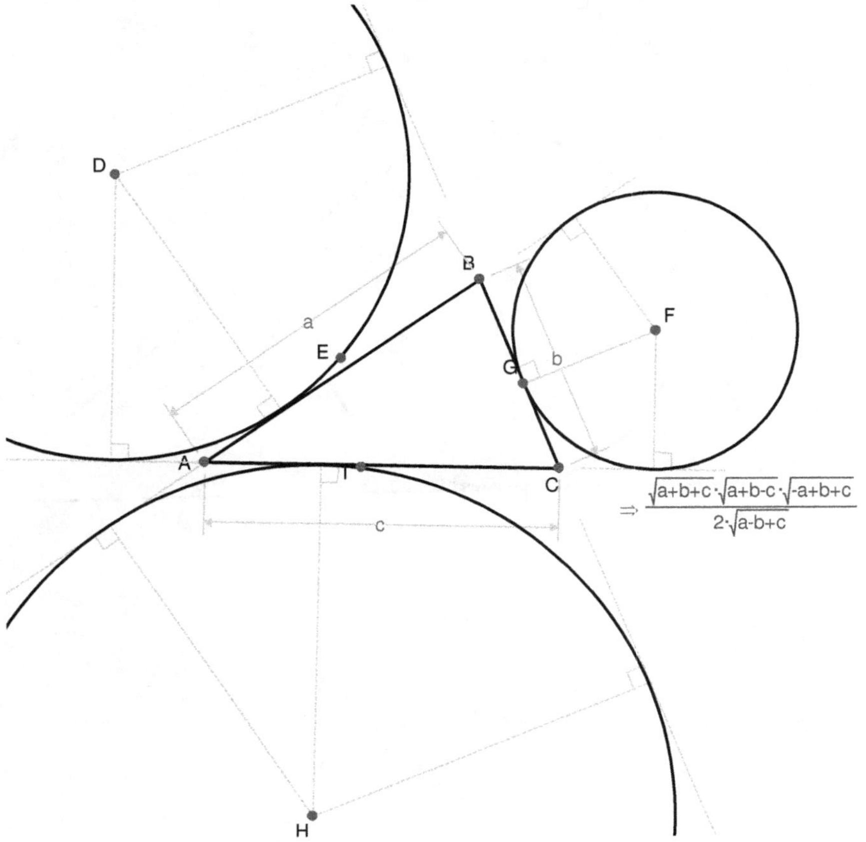

What is the ratio between this radius and the area of the triangle?

Can you prove the result using areas of triangles ABF, ACF, BCF?

What are the radii of the other two excircles?

What is the product of the radii of the three excircles and the incircle?

We examine the triangle joining the centers of the excircles:

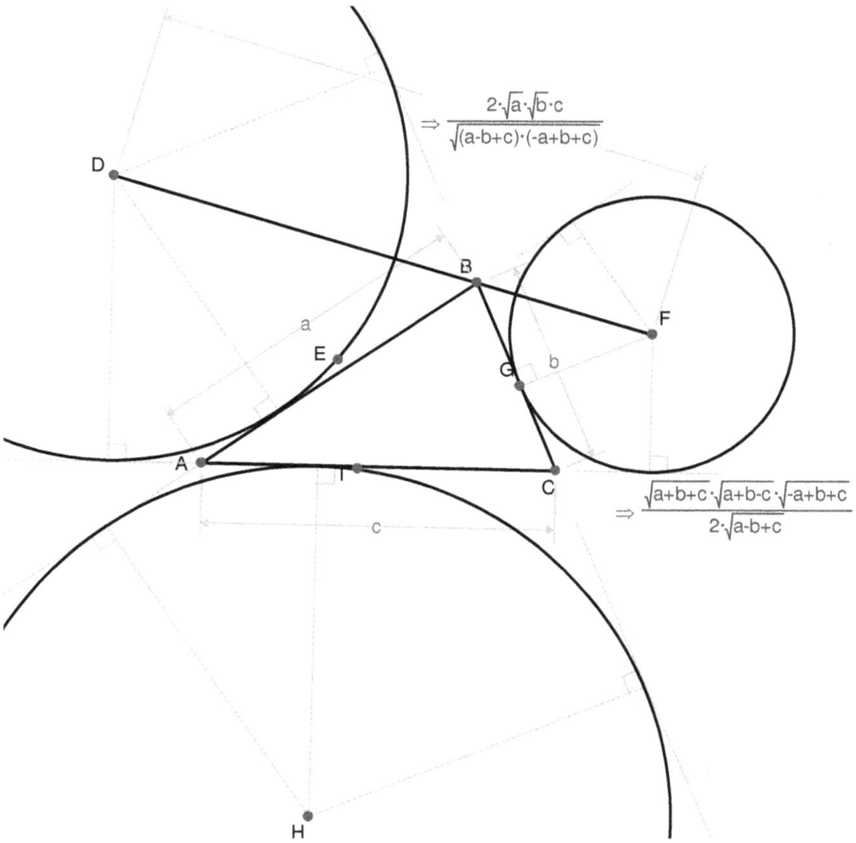

Can you prove that B lies on DF (think in terms of symmetry)?

Can you prove the length?

www.ingramcontent.com/pod-product-compliance
Lightning Source LLC
Chambersburg PA
CBHW080545220326
41599CB00032B/6364
* 9 7 8 1 8 8 2 5 6 4 1 2 5 *